Environmental Stress Screening Handbook

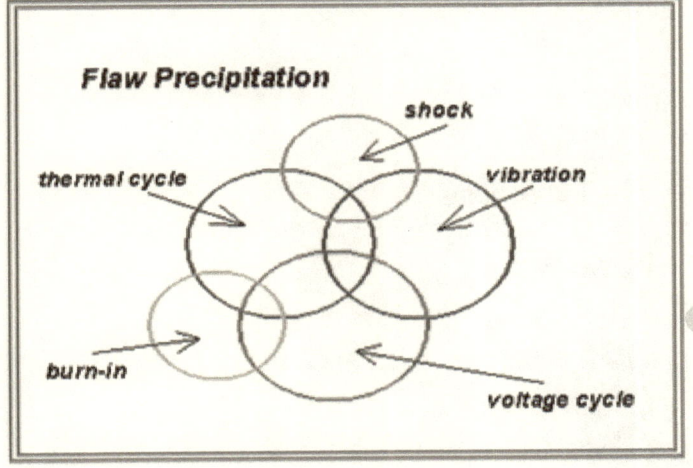

Flaw Precipitation

shock

vibration

thermal cycle

burn-in

voltage cycle

Stop The Guessing

John J. Quinn

authorHOUSE

1663 LIBERTY DRIVE, SUITE 200
BLOOMINGTON, INDIANA 47403
(800) 839-8640
www.authorhouse.com

First published by AuthorHouse 08/11/04

ISBN: 1-4184-6169-5 (e)
ISBN: 1-4184-2819-1 (sc)

Library of Congress Control Number: 2004105067

Printed in the United States of America
Bloomington, Indiana

This book is printed on acid-free paper.

Preface

This book describes the analytical approach to solving the development / comparison of Environmental Stress Screens, It is modeled from DOD-HNBK-344 and 344A which was developed by Rome Air Development Center (RADC). The initial Technical Reports (TR) were released in 1981, and the handbook was released in 1986. The handbook is divided into five procedures that cover defect density (defect numbers assigned to parts, PWA's and connections). Screen selection and placement (calculations for thermal & vibration screens). Failure Free Acceptance test (calculates the failure free time during ESS from the MBTF/yield requirements). Cost Effective Analysis (allows the user to evaluate a total screen using real dollars). Monitoring, Evaluation and Control (control charts to monitor the dropout/failures at each screen level).

Section 1 is entitled "An Introduction of How DOD-HNBK-344" can influence new and existing Environmental Stress Screening Programs. It describes the math models developed to calculate thermal and vibration screens, explains the use of a defect density worksheet, compares two existing screens and calculates a failure Free Acceptance Test for a sample screen. The benefits of an ESS Program are also described. The Precipitation Efficiency equations are also contained in this section.

Section 2 describes an Environmental Stress Screen Model. This ESS Model can be used as a foundation to develop new product screens for development and production. The model was developed to existing Programs with the lowest level screen applied to circuit cards and the highest level screen applied to section/Forebody.

Section 3 uses the information in the previous sections and shows how this information can be used to develop a computer spreadsheet program such as EXCEL to develop this ESS model. It describes how this spreadsheet program automatically performs the math equations required for test strength, fallout and defects. This section also includes cost analysis and control worksheet examples.

Section 4 entitled "Lesson Plan/Study Guide" describes a baseline & introductory Environmental Stress Screening Course. All viewgraphs are included in this section.

Section 5 is an example of the use of a spread sheet to develop a screening program. The topics discussed are temperature cycling, random vibration, screen composites (a value for more than one screen) and damage index (how much of the unit's life has been used due to screening).

Section 6 contains Excel worksheets for ESS calculations: temperature cycling, constant temperature (burn in), random vibration, swept sine vibration and single frequency sine vibration.

Appendix A is an explanation of EXCEL worksheets to calculate Screening Strengths, Screening Composites and "G rms" to calculate Vibration equations.

Appendix B contains definitions of common ESS terms.

Appendix C contains an explanation of a Failure Review and Corrective Action System (FRACAS) and Screening Review Board (SRB). Both of these items are required for environmental screening success. If these items are not in place, the screening program will become a hidden factory.

Finally there is a way to compare apples and oranges (sine vibration vs. random vibration vs. thermal cycling vs. constant temperature). Every procedure concludes with a real number that has quantitative meaning. No longer do we have to guess and experiment to develop an ESS program, or how it fits in with all the various testing that is required. ESS is defined as a process not a test. ESS is used to stimulate latent defects to the detectable state and not to simulate the end environment. New program development should follow "The Steps to Success" illustration. A new program has environmental requirements as a given and must meet field requirements. Engineering uses these requirements to select the parts for designing the "Unit Under Test" (UUT). Parts screening can be a mini ESS, only new or known problem parts would require 100% screening, all others would be controlled via an audit.

Environmental Stress Screening weeds out the latent defects due to part selection and workmanship during the assembly of the UUT. Acceptance testing is proof that the UUT meets its performance requirements and is manufactured the same as previous UUT. Qualification Tests are performed to prove that the UUT can withstand one lifetime of the end environments, from manufacturing through shipment and finally its mission. Test Analyze and Fix is an on-going test to verify MBTF and product wear out. If the testing limits are followed as shown in the "Steps to Success" figure below, the outcome will be a reliable product.

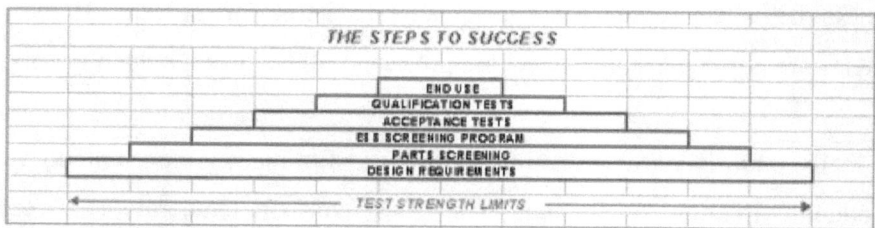

THE STEPS TO SUCCESS

END USE
QUALIFICATION TESTS
ACCEPTANCE TESTS
EI S SCREENING PROGRAM
PARTS SCREENING
DESIGN REQUIREMENTS

TEST STRENGTH LIMITS

In the near future there will be an electronic version of this document to be used with a computer.

Table of Contents

Table of Figures

1.1 INTRODUCTION and PURPOSE

This section is divided into two (2) parts, Introduction (description of Handbook) and Purpose (benefits of ESS).

The first section provides a brief description of DOD-HNBK-344 and how it should be used to compare Environmental Stress Screens (ESS) from program to program or product to product. It provides examples from the 5 major sections of the DOD handbook.

1) Part faction defective – How to calculate a defect as a function of part quality.

2) Screen Selection and Placement – How to select a more cost effective screen.

3) Failure Free Acceptance Test (FFAT) – The last portion of the ESS can be used as an Acceptance test (reduction in test time).

4) Cost Effective Analysis – How to calculate the screening costs.

5) Monitoring, Evaluation and Control – Failure Analysis must be performed and corrective action must be taken.

The second section relates to the benefits if DOD-HNBK-344 is used as a guideline. The benefits are:

1) Screen comparison.

2) Failure Free Acceptance Tests.

3) Future Request for quotes (RFQ).

4) Dovetails with TQM & SPC.

5) Cost Effectiveness.

Environmental Stress Screening is not new. RADC published its first technical report TR 87-81 in 1981. This document was entitled "BURN-IN: WHICH ENVIRONMENTAL STRESS SCREENS SHOULD BE USED". It was the first attempt to introduce math models. Since then new ideas have constantly surfaced such as DOD-HNBK-344 & 344A. As more test data is available the mathematical models will possibly change. In 1986 DOD-HNBK-344 was published and revised the math models of TR 87-81 because of new test data. LITTON of Canada was tasked to evaluate DOD-HNBK-344 using three different product manufacturing lines. The results did not change the basic math models. New environments may be introduced, as an example Temperature Shock has been addressed in the handbook. However, can this be used in ESS? How is it calculated? Its cost effectiveness must be addressed. Temperature Shock requires two chambers or a chamber with an elevator apparatus to change from one temperature to another. The rate of temperature change is far greater than thermocycling and the phenomenon of solder creep can not be detected. The implementation and evaluation of ESS should be an on-going task to improve the product reliability and corporate profits.

The purpose of this report is to standardize the approach in developing ESS programs. This standardized technique will be used on all new programs and products and supports the concept of Total Quality Management (TQM) & Statistical Process Control (SPC) because its foundation is based on quantitative solutions. This technique is not intended to standardize actual ESS programs or ignore all previous ESS experience. This technique will standardize the approach used to develop ESS programs while incorporating past experience to tailor the new program. All Programs/Products have different screens for various reasons such as yield/MBTF, physical characteristics, manufacturing processes and parts. Throughout this report all deliverable milestones will be driven by the techniques of DOD-HNBK-344.

1.2 BRIEF DESCRIPTION OF DOD-HNBK-344:

DOD-HNBK-344 is a quantitative approach to develop a cost effective ESS which includes planning, monitoring and control. There are five sections in the handbook.

1) Part Fraction Defective – (procedure A in the handbook) – Part Fraction Defective (PFD) is a means to determine what the total number of defects that are on a particular unit under test (UUT). PFD is influenced by component type (transistors, resistors, capacitors, etc.), the quantity of that component(s) and the quality level (b-1, jtx, m, etc.). Another factor is the type of connections that the UUT is manufactured to (hand solder, crimp, weld, etc.). MIL-HNBK-217 contains this type of information and was used to model the defect tables that are found in DOD-HNBK-344. Now with prior knowledge of the component type , quality level and the quantity

John J. Quinn

of the part or connection type, a matrix can be prepared with the total number of defects for the UUT. An example of this can be found in figure 1. The part type column lists all the different electronic devices, printed wiring boards and the type of connection used to manufacture the UUT. The quality level column shoes the rated quality level of the device. An explanation of rated quality level can be found in MIL-STD-217 "Reliability Prediction of Electronic Equipment". The part fraction defective column lists the estimated defect that the part contains and the last column estimated defects is the total defects per 106 for that device. For example the resistors used on figure 1 has a quality rating ER-M and 18 resistors are used on this UUT. Multiplying fraction defective of 23.8 by 18 yields a total estimated number of defects per 106 of 428.4. Therefore the total estimated defects of figure 1 is 6160.9. This is based on the quality of the part used and the quality of the part. The defect tables can be found in DOD-HNBK-344. This type of information for example can be calculated when a reliability study is underway for a new development program. The UUT can be a discrete component, circuit card assembly, sub-assembly, section or an end item.

DEFECT ESTIMATION WORKSHEET EQUIPMENT: PROCESSOR UNIT - 9664009		Prepared By *JJ Qm*	Date 6/20/91	
Part Type	Quality Level	Qty	Fraction Defective	Estimated Defects
Microelectronics	B-O	49	87	4263
Transistors				
Diodes	JANTX	1	46.9	46.9
Resistors	ER-M	18	23.8	428.4
Capacitors	ER-M	1	115.3	115.3
Inductive Devices				
Rotating Devices				
Relays				
Switches				
Connectors	M/S	1	168	168
Printed Wiring Boards	M/S	1	1139.3	1139.3
Connections, Hand Solder				
Connections, Crimp				
Connections, Weld				
Connections, Solderless Wrap				
Connections, Wrapped & Soldered				
Connections, Clip Termination				
Connections, Reflow Solder				
Total				6160.9

Figure 1 – Defect Estimation

2) Screen Selection and Placement – Procedure B in DOD HNBK -344 – This process selects the most cost effective screens at various levels of the manufacturing process. Some important factors which influence the screen selection are:

a) Incoming defect density; the quality and quantity of the parts used in the design of the UUT. This is referred to as Defects In (D_{in}).

b) Screen selection; the selection of thermocycling and /or random vibration, refer to Figure 1-2 for Assembly Defect Types precipitated by thermal and vibration screens. Reference RADC TR 82-87

Defect Type	Thermal Screen	Vibration Screen
Defective Part	X	X
Broken Part	X	X
Improperly Installed Part	X	X
Solder Connection	X	X
PCB Etch – shorts & opens	X	X
Loose Contact		X
Wire Insulation	X	
Loose Wire Termination	X	X
Improper Crimp	X	
Contamination	X	
Debris		X
Loose Hardware		X
Chafed /pinched wires		X
Parameter Drift	X	
Hermetic Seal Failure	X	
Adjacent Boards/Parts Shorting		X

Figure 2 – ASSEMBLY DEFECTS

c) Test Equipment; how accurate is the test equipment ability to detect latent failures. Latent Defect – An inherent or induced weakness, not detectable by ordinary means, which will either be precipitated to early failure under environmental stress screening conditions or eventually fail in the intended use environment (definition of latent defect from DOD-HNBK-344). This is referred to as Probability of detection (P_d), and is used to calculate Screening Strength (SS) and Defects out (D_{out}). Where Dout is the defects passed on to the next higher level of assembly.

$F = D_{in} \ X \ SS \ where \ SS = P_e \ X \ P_d$ (Equation 1)

$D_{out} = D_{in} - (\ 1 - SS \ )$ (Equation 2)

F = fallout, SS = screening strength, P_e = Precipitation efficiency (This term was changed in DOD-HNBK-344A. It was SS in 344).

Terminology changes:

DOD-HNBK-344	DOD-HNBK-344A
Test Strength (TS) = SS X P_d	SS = P_e X P_d

d) Design limits; The ESS environmental stress limits should not exceed the design environmental stress limits or the limits of the parts used to manufacture the UUT.

An example of calculating P_e can be shown using figure1-3. In this example the screens for module level, plate or major assembly and sections for two product lines (Product A & Product B) will be compared. Product A at the module level has 15 thermal cycles with a range from $- 50^{\circ}C$ to +

7

85°C and a ramp of 15°C per minute. Using the Precipitation Efficiency Factors Table (5-15) from DOD-HNBK-344A a Pe of .9999 is determined. It should be noted from the table that by decreasing the number of cycles from 15 to 12 provides the same Pe. This will also decrease the cost of the screen (probably about 4 hours). Product B at module level has 5 thermal cycles with a range from – 40°C to + 85°C with a ramp of 15°C per minute. This time the thermocycle equation will be used.

Pe, thermocycle (Pe_{tc})

$$Pe_{tc} = 1-\exp\{-.0017\ (r + 0.6\ )^{0.6}\ [\ \ln (e + dt)]^3\ (N_{cyc})\}$$ (Equation 3) where r = temperature range (°C), dt = ramp in °C/min. and N_{cyc} = number of thermal cycles.

Solving the equation for product B, ($Petc$ = .965). Both products A & B exhibit a very high Precipitation Efficiency at the module level. This is the most cost effective method, a high Precipitation Efficiency at the lowest level of assembly. As the product is assembled the cost to repair increases by an order of magnitude at the next higher level of assembly, for example:

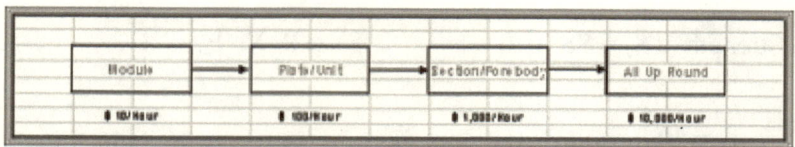

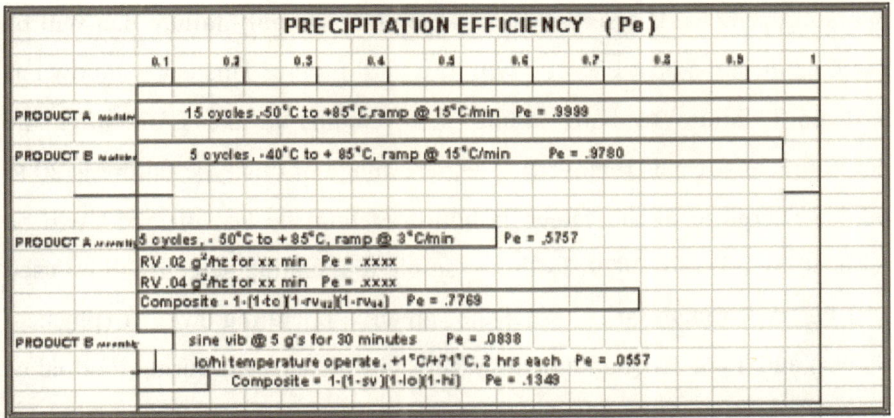

Figure 3 – Precipitation Efficiency

Product A at the Plate level has both random vibration and thermocycling applied. Five thermal cycles with a range from − 50°C to + 85°C and a ramp of 3°C/min. will be applied. The 3°C/min is below the minimum recommended ramp of 5°C/min. The large mass of Product A influences the actual ramp. Several things could be implemented to increase the ramp of Product A:

1) Increase ramp by overshooting the temperature setting of the thermal chamber. Thermal couples must be used to ensure product does not go above high temperature. Recommend a thermal screen be performed to determine temperature profile.

2) Increase the thermal chamber air speed. This will decrease the thermal lag.

3) If Product A has removable covers, remove them. This will reduce thermal lag of Product A.

In this example dt is 3, r is 135 and Ncyc is 5. Using equation 3 again yields a Pe_{tc} = .5757. This is a good time to introduce operating and non-operating screens. If a screen is non-operating the only failure that will be detected during a post ESS test will be a hard failure. This is also known as a Type I failure (a hard failure that remains in the UUT after the environment stress is removed). The percentage of this type failure is 20 to 50 %. Therefore, if the screen is non operating the Pe_{tc} of .5757 will be reduced by 50% to .2879. In this example a operating screen will be used and Pe_{tc} will remain at .5757. If the screen is operating Type II (physical defect) and Type III (Functional defect) will be detected during thermal or vibration stresses are applied. Therefore, if a screen is operational all latent defects should be detected.

Looking at the random vibration at Product A plate level shows that 2 vibrations occur, one at an ASD of .02 g^2/hz and the other at .04 g^2/hz. The Precipitation Efficiency calculations are as follows:

To convert .02 g2/hz to g's rms the square root of the area under the curve must be calculated. The frequency range is 550 hz.

G's rms = $((ASD) X frequency range))^{1/2}$ (Equation 4)

G's rms = $((.02) X (550))^{1/2}$

G's rms = 3.31

Using the random vibration formula for Precipitation Efficiency for 4 minutes of vibration yields:

Pe_{rv1} = 1-exp[-.0046 $(g)^{1.71}$ (t)] (Equation 5)

where g = 3.31 and t = time in minutes (4).

Pe_{rv1} = 1-exp[-.0046 $(3.31)^{1.71}$ (4)]

Pe_{rv1} = .1328

2) To convert .04 g2/hz using equation 4 yields:

$$G's \ rms = ((.04) \ X \ (550))^{1/2}$$
$$G's \ rms = \ 4.6$$

Using the random vibration formula for Precipitation Efficiency for 8 minutes of vibration and frequency range of 550 hz yields:

$$Pe_{rv2} = 1\text{-}exp[-.0046 \ (4.6)^{1.71} \ (8)]$$
$$Pe_{rv2} = .3936$$

The composite of the two vibrations can be calculated as follows:

$$Vib \ Comp = 1 - (1 - pe_{rv1}) \ (1 \ - \ Pe_{rv2}) \quad (Equation \ 6)$$
$$= 1 - (1 - .1328) \ (1 - .3936)$$
$$= .4741$$

Therefore the total composite Pe for Product A at the module level for 5 thermal cycles and 2 random vibrations is .7769.

$$Pe_{comp} = 1\text{-} \ (\ 1 - thermal \ Pe \ ) \ (\ 1 - Vib \ Composite \ )$$

Examining Product B shows that 30 minutes of sine vibration at 5 g's yields a (Pe) of 0.0838, from the (Pe) of DOD-HNBK-344 table 5-16. It should be noted that from all the data published about vibration, sine vibration is about 1/10 effective as random vibration. Single Frequency or swept sine vibration only excites one frequency at a time where random

John J. Quinn

vibration excites all the frequencies simultaneously. Refer to Figures 1-4, 6, 7 & 8. Low and high temperature operate are constant temperature screens (burn-in). Constant temperature screens are not as effective as thermal cycling.

Constant Temperature Screen:

$$Pe_{ct} = 1 - \exp[-.0017(r + .6)^{0.6}(t)] \qquad \text{(Equation 7)}$$

where r = temperature range above ambient in °C & t = time in hours.

High temperature at + 71°C;

Range = 71 − 25 = 46, t = 2 hrs (25°C is ambient temperature)

$$Pe_{ct} = 1 - \exp[-.0017(46 + .6)^{0.6}(2)]$$

$$Pe_{ct} = .03351$$

Low temperature at + 1°C;

$$Pe_{ct} = 1 - \exp[-.0017(24 + .6)^{0.6}(2)]$$

$$Pe_{ct} = .0230$$

The composite of both constant temperature screens calculates to be .0557. This is a weak screen as compared to Product A's 5 thermal cycles (.5757). The total composite for Product B at the module level is (.1349). Using Figure 2 it can be seen that Product A has a high (Pe) and will be successful in detecting latent failures. Product B is not as effective, and will not be successful in detecting latent failures.

3) Failure Free Acceptance Test (FFAT). Procedure C/DOD-HNBK-344

FFAT is defined as the time interval during which the UUT must operate without failure, while exposed to a particular environmental stress. The time interval is a function of the yield or MBTF (Figure 1-11) requirements. At the system level this FFAT could be used as the official Acceptance Test. Here is another cost savings due to the reduction of test time. An example of calculating FFAT is as follows:

o Failure Free Test Period (given the following)

       - MBTF of 500 hours

       - Yield = .51 (from Figure 1-11 and Table 2.27 DOD-HNBK-344)

o Failure Free Period = N_g

       N_g = 1/MBTF = 1/500 = .002

       f = Failure Rate from Pe table Fig 5 = .7852 (Ramp 15°C & Range

140)

       Failure Rate Ratio = f/N_g = .7852/,002 = 392.6

       Using Figure 1-5, Fail Time T_f = 1.5 hours

       Failure Free Time = 1.5/0.7852 = 1.91 hours

This is for a yield of .51 with a 90 % lower confidence bound, the product/UUT must operate 2 hours Failure Free during the Stress Screen.

4) Cost Effectiveness Analysis, Procedure D in DOD-HNBK-344

Costs are determined at each level of assembly screen. These costs include:

       a) Fixed Screening Costs – facilities, test equipment, test fixtures and procedures/training.

  b) Variable Screening Costs – labor, management, failure analysis and data management.

  c) Average Cost to Repair – actual manufacturing costs for rework and repair at each level of assembly, if not known an approximate value is given on the worksheet (figure 1-9).

  NOTE: If the cost per defect is greater than the threshold cost, Cd > Ct the screen is not cost effective and should be re-evaluated.

5) <u>Monitoring, Evaluation and Control</u>, (Procedure E in DOD-HNBK-344)

There are several procedures listed in DOD-HNBK-344 that describe how to manage the control of the screens at various assembly levels. Figure 1-10 – Comparison of Actual vs. Planned Defects is a good example on managing a screen. A computer spreadsheet using CCALC Plus/Excel will be developed to calculate the control charts for each screen. For training purposes to screen will be changed to demonstrate how control charts are automatically recalculated.

1.3 BENEFITS

A numerical value can be determined for a particular screen by following the steps outlined in Section B of DOD-HNBK-344. This value is referred to as Precipitation Efficiency (Pe) and is found in the tables in sections 2 & 3. This will allow a numerical comparison between screens used in a particular program or other programs and will eliminate the need for trial and error techniques used to develop new product stress screens. A comparison between thermocycling and vibration can also be made. An example of this comparison has been explained using figure 3 (Product A vs. Product B).

2) Failure Free Acceptance Time – the time that the UUT must operate during the ESS (usually at system level) to assure that the Defect Goals have been achieved. Arbitrary selection has no quantitative assurances. If a failure appears during FFAT it is repaired and FFAT starts from the beginning. The length of FFAT is a function of the yield requirements, type of screen and statistical confidence that the yield has been achieved. It can be used as a measure of system level screen effectiveness or as part of the actual Acceptance Test. This can eliminate costly test time by proposing that the Acceptance Test of a particular UUT (especially spares) be part of the ESS Plan.

3) Future RFQ's that requires ESS in accordance with DOD-HNBK-344 would be advantageous. An exercise in bidding has already been performed using the complexity of a Missile Program as a model. Task Descriptions, Cost Rationales and 1037's were prepared for all 5 Procedures of DOD-HNBK-344.

o Procedure A – Parts Fraction Defective Estimate........... 20.5 MM
o Procedure B – Screen Selection .... 1.0 MM
o Procedure C – Failure Free Acceptance Test .... 1.0 MM
o Procedure D Cost Effective Analysis .... 0.5 MM
o Procedure E – Monitoring, Evaluation & Control .... 2.5 MM

(this is for the initial development of the control charts. The actual monitoring and data collection for failure analysis is an on-going effort and should be bid separately).

4) Dovetails with Total Quality Management (TQM) – ESS IAW DOD-HNBK-344 can satisfy many of the principles of TQM for example:

o Customer Satisfaction – ESS can produce significant improvements in reliability and reduce field failure costs. A cost effective ESS can impact the quality and reliability of electronic products delivered to the Government. ESS is interrelated with the requirements of MIL-Q-9858A/ISO 9000 and MIL-STD-785 method 301.

o Individual Participation – ESS at each level of assembly (CCA's, Plate/Units and Section/System) eliminate <u>latent</u> failures and pass to the next higher level ESS as a higher quality UUT.

o Continuous Improvement – ESS is a dynamic process and changes as the product reaches its yield goals or where failure analysis dictates.

o Robust Design – ESS during the development phase can be used to aid in solving design problems. This could be as high as 10 % during initial development.

o Variability Reduction – ESS achieves this by refining the manufacturing process through test and analysis.

o Statistical Thinking – Process measurement through ESS control charts at each screen level can provide data for possible problem avoidance and quality improvements.

5) Cost Effectiveness Screens – Screening at the lowest possible level is cost effective in terms of troubleshooting failures and rework (eliminating the "Hidden Factory"). Troubleshooting and repair at the next higher level of assembly increases cost at least by an order of magnitude. Cost savings to Raytheon and the Government is through improved reliability and a reduction of field repair costs. Screening at the parts level is often the most cost effective way to eliminate defects prior to CCA assembly. However, 100 % re-screening may not be cost effective due to the increased failures due to:

o Handling causing physical damage

o Handling due to Electrostatic Discharge (ESD)

o Seller and Buyer testing inconsistencies

An example of this is the comparison of Northrop's Inertial Instrument Unit screen. The old screen was 30 minutes of sine vibration and a high temperature soak. The new screen is 10 minutes of sine vibration using a triax vibration fixture that allows 3 axis of vibration at one time. The UUT sits 45 degrees off the prime input axis.

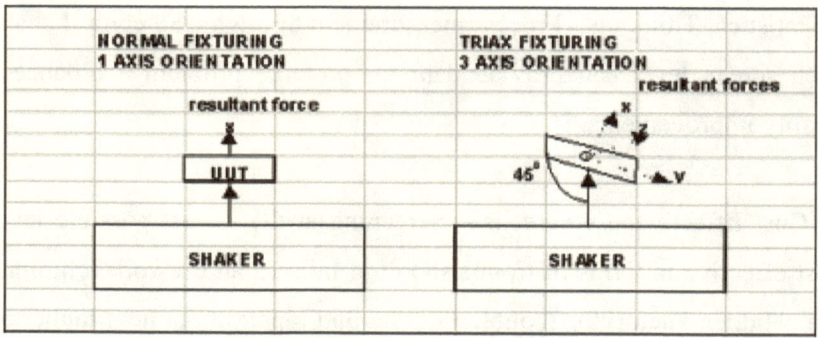

The new screen went to 3 thermal cycles instead of the high temperature soak. Calculating the Pe of both new and old screens indicate the vibrations were a wash out, both methods had a low Pe and the fallout data from the screens supported that fact. By adding 3 thermal cycles in place of the high temperature soak the new screen became 6 times stronger than the old screen. This was also supported by the fallout data showing an increase of 60 % failures are due to thermocycling.

1.4 PRECIPITATION EFFICIENCY EQUATIONS REVIEW

Thermocycling

$$Pe_{tc} = 1 - \exp \{ -.0017 (R + 0.6)^{0.6} [\ln (e + DT)]^3 (N_{cyc})\}$$

where DT = rate of change (ramp) in degrees C

R = temperature range (hi to lo) in degrees C

Ncyc = number of thermal cycles

Constant Temperature (Soak)

$$Pe_{ct} = 1 - \exp [-.0017 (R + 0.6)^{0.6} (T)]$$

where R = temperature range in degrees C, hi temp is above ambient of 25°C.

T = time in hours

Random Vibration

$$Pe_{rv} = 1 - \exp [-.0046 (G)^{1.71} (T)]$$

where G = g's rms and T = time in minutes

Sine Vibration (Swept Sine)

$$Pe_{ss} = 1 - \exp [-.000727 (G)^{0.853} (T)]$$

where G = g's and T = time in minutes

Sine Vibration (Single Frequency)

$$Pe_{sf} = 1 - \exp [-.00047 (G)^{0.49} (T)]$$

where G = g's and T = time in minutes

SECTION 2

2.1 ENVIRONMENTAL STRESS SCREENING MODEL

The Environmental Stress Screening (ESS) Model was developed from DOD-HNBK-344 & 344A and is divided into 3 levels of screens.

1) Circuit card level – 12 thermal cycles

2) Plate/Unit or major assembly level – 6 thermal cycles and 10 minutes of random vibration per axis at .04 g^2/hz.

3) Section /Forebody level – 3 thermal cycles and 10 minutes of random vibration per axis at .02 g^2/hz.

The environments used to precipitate the latent failures are 1) Thermocycling with a range from – 45°C to + 85°C and a ramp of 15°C per minute. 2) Random Vibration using the standard NAVMAT P-9492 profile for 10 minutes per axis, refer to Figure 4.

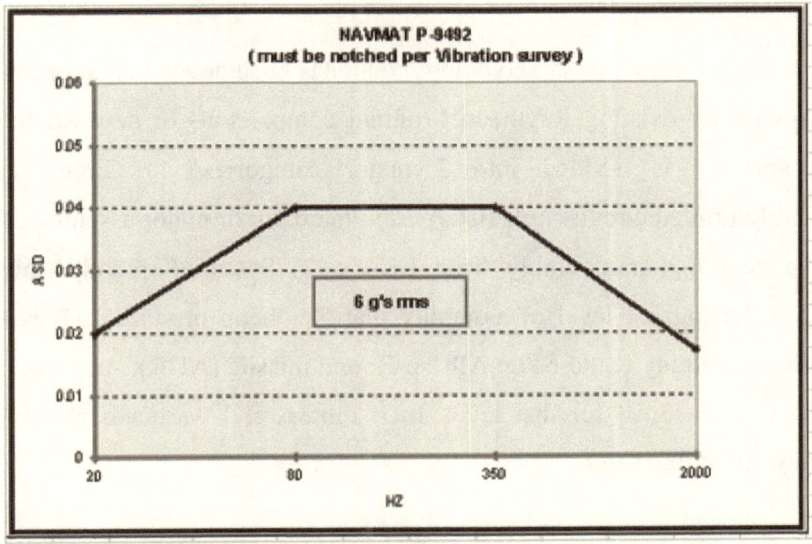

Figure 4 – NAVMAT

A very important part of each screen is the Failure Analysis that must be performed to eliminate the HIDDEN FACTORY by corrective action. The most cost effective screen is at the lowest level of assembly which in the ESS model is the circuit card level. Again the cost to repair is a factor of ten as the screen level increases from circuit card to plate to section level.

The precipitation efficiency for each screen has been calculated from DOD-HNBK-344 and the fallout is predicted as a function of the defect density. The defect density is a value used from tables in DOD-HNBK-344 and MIL-STD-217. These values are derived from the Quality and Quantity of the components used to manufacture the Unit Under Test (UUT).

"QUALITY PARTS + STRONG ESS = HIGH RELIABILITY"

The Environmental Stress Screening model has been developed to establish a baseline for existing Raytheon Program comparisons or new Raytheon programs. It is divided into 3 major categories: 1) Circuit card assembly/printed circuit cards (CCA). 2) Plate/Unit or major assembly, this is the next higher assembly from CCA's. 3) Section/Forebody, this is usually the highest level of assembly that Raytheon produces. However, the final assembly could be an All-Up-Round missile (AUR). And a screen could be developed for that level. Inert motors and warheads should be considered at this level.

Looking at the CCA section (Screen # 1) – This model uses 12 thermal cycles with a range from – 45°C to + 85°C with a ramp of 15°C/minute. This yields a Precipitation Efficiency (Pe) of .9999. This Pe can be calculated with the equation shown on figure 2-1 or can be looked up in the tables contained in DOD-HNBK-344 or 344A. The fallout from screen # 1 (FD_1) that is calculated for screen # 1 is the Defect Density (Din 1) X Pe X Pd, where Pd is the Probability of detection (est equipment capability). There are tables in DOD-HNBK-344/344A that contain this information. In this model the worst case of .85 is used. Therefore the fallout can be calculated as:

$$FO_1 = D_{in\ 1} \text{ X Pe X Pd}$$
$$FO_1 = D_{in\ 1} \text{ X .9999 X .85}$$

Din 1 is unknown, therefore 85% of the latent defects should be detected. Each CCA type will have it's own (defect density in) number. The initial failure rates for parts and connections can be found in MIL-HNBK-217D, notice 1 and DOD-HNBK-344/344A which contain a defect density worksheet example. Now for this example, if the total estimate was 3.0 and using screen # 1, the expected fallout would be 3.0 X .8499. The fallout is 2.5 defects detected, the remaining 0.5 is carried into the next higher assembly as part of $D_{in\,2}$.

The fallout from screen # 1 has to go through Failure Analysis. The first determination is the type of failure. A <u>Patent</u> failure is one that is detected by inspection or functional test. A <u>Latent</u> failure is one that is detected via an environmental stress applied to the UUT. After the classification of the failure the quantity should be examined to see if it falls within limits of the control charts. The control charts could use the predicted fallout, +/- 3 sigma limits. If the results are within the control chart parameters the present screen is adequate. However, if the results are out of the control chart parameters a flag should go up and an analysis of the screen should be performed. The fallout data and failure analysis is just as important as the ESS. The fallout data should dictate the dynamics of the screen, At some point in time if the fallout data is acceptable the 100 % screen requirement should be reduced to an audit percentage.

The last remaining subject of screen #1 is the type of environments to use for the screen. In this example thermocycling is recommended. The thermocycle profile goes hot first, this will dry out the UUT and not cause an induced failure at low temperature due to moisture in the UUT (refer to Figure 5). At this point in the manufacturing cycle the CCA will not have been conformal coated.

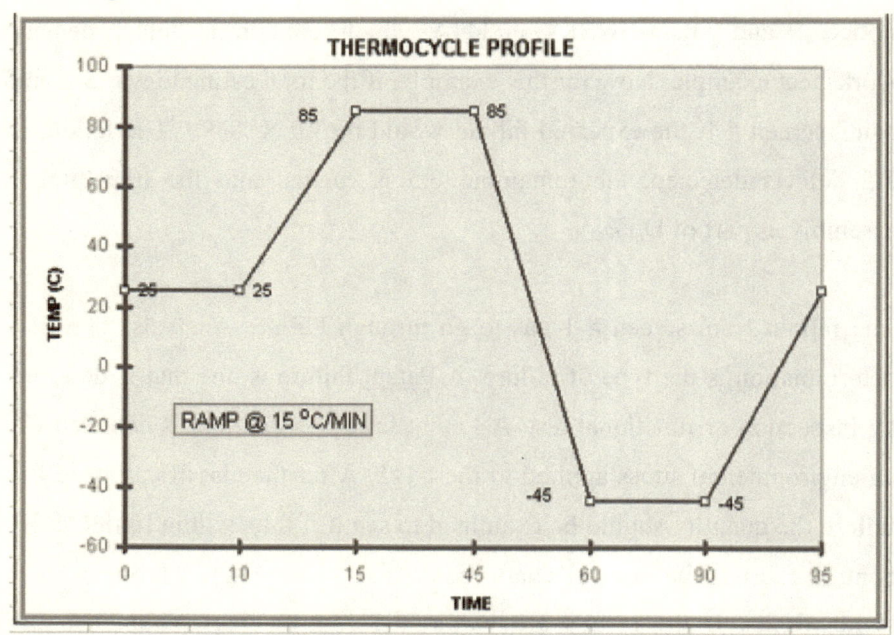

Figure 5 – Thermal Cycle Profile

12 thermal cycles (operating) have been chosen with a range from – 45°C to + 85°C and a ramp of 15°C/minute. This yields a Pe of .9999. Using a high Pe at the CCA level is most cost effective. Here is another example of using the fallout data, if the fallout data shows that all the failures are occurring during thermal cycles 1 through 4, the screen should be revised to possibly 6 thermal cycles instead of 12. This is a large cost savings. Save all fallout data and analysis to convince the customer of the screen change.

There are many reasons during thermocycling to have power on and monitor the UUT electrical parameters:

1) Applying power during the high temperature transition will help the UUT to reach the high temperature limit at a faster ramp rather to rely on

24

the chamber's ability to reach high temperature limit. Another method to increase temperature ramp is to over shoot the high temperature setting in the chamber, ie if the high temperature limit of the UUT is 85°C – set the chamber temperature to 100°C. Thermocouples must be used in this technique to ensure that the UUT does not go above 85°C. This is part of the thermal survey to determine thermal chamber set up. If the thermal chamber air flow can be increased, this would decrease the ramp time. The faster the air flow the quicker the UUT will achieve high temperature. If the UUT has removable covers they can be removed during temperature ESS. The UUT does not have to be configured as in acceptance or qualification testing. All of the above method to decrease the ramp time is accomplished by decreasing thermal lag.

2) Monitoring (power on & electrically tested) will allow to know what point in time for vibration screens failures and during which thermocycle the UUT failed. This information can be used to alter/revise the current screen.

3) During ESS intermittent failures can be detected as well as out of specification conditions. For example during high temperature an output signal drifts out of tolerance and when the temperature is lowered the signal will within specification. Unless the UUT is monitored during the thermal stress this failure condition would go undetected.

4) Failure Free Acceptance Tests (FFAT) can be performed and possibly replace the actual Acceptance Test Plan and Procedure. This would reduce test time.

5) Monitoring during ESS can eliminate soma pre and post testing. If the UUT has passed ESS at that level there is no need to test again at ambient (post test).

6) Voltage margins and on/off (voltage snaps) conditions can be included in the screen, thereby eliminating additional testing.

Vibration has not been mentioned for CCA screening. Here's where experience comes in, if you use high profile parts (relays, IC cans) or parts with a history of failing during vibration (this data is available at incoming parts inspection) then vibration could be added to screen #1. For this model the vibration screen will be applied at the next higher level of assembly (plate/unit level).

At the next higher level of assembly, screen #2 will be used. This screen will utilize 6 thermal cycles and 30 minutes of random vibration (10 minutes per axis). The (Pe) can be calculated or looked up in the tables of DOD-HNBK-344/344A.

The total composite (Pe) for this screen is:

Comp = 1 – (1 – vibration Pe) (1 – thermal Pe)

Comp = 1 – (1 - .9840) (1 - .9864)

Comp = .9995

The defect density for this screen is in the form of 1) – The remaining defects passed on from screen #1 and 2) – Any purchased parts/assemblies (gyro's, acceleromenters, etc) and cables. It should be noted that cables are recommended for a separate screen. Cables have a high failure rate during environmental testing and some Programs have cable ESS requirements. Vibration will be performed first, then thermocycling.

The vibration screen is quicker in test time and should precede the temperature screen. If a failure occurs during vibration consider it a cost saving due to test time. Experience has shown that vibration can loosen connections to a point of almost failing (solder joints) and then the expansion and contraction of thermocycling brings out the failure mode. A good rule of thumb "SHAKE & BAKE" at any level of assembly higher than CCA's. The vibration profile used for this screen is from NAVMAT P-9492 and has proven to surface latent failures and not overstress the UUT. The thermocycle limits remain the same as screen #1, - 45°C to + 85°C.

1) If the UUT temperature specifications at this point of assembly are lower than the screen temperature, use the screen temperature limit. ESS environments do not have to be less severe. A good ESS is developed to Stimulate failures not Simulate the end item environments. If the UUT has a component that has a proven history of high temperature problems then the screen temperature should be adjusted for this UUT only.

2) If the UUT has a large mass it may be difficult to maintain a 15°C/minute ramp. Several options have to be considered:

    a) Purchase a high quality chamber that can satisfy the ramp requirements of CCA's, Plates and Sections. This would be a high cost item but may result in a large payback due to less test time required to reach the same (Pe) requirement.

b) Reduce the (Pe) for thermocycling due to ramp requirements dictated by the mass of the UUT and thermal chamber. Using a 5°C/minute ramp will reduce the (Pe) to .7864 from .9864. This makes the composite for screen # 2 equal:

$$Pe_{comp} = 1 - (1 - Pe_{vib})(1 - Pe_{therm})$$
$$Pe_{comp} = 1 - (1 - .9840)(1 - .7864) = .9966$$

Multiplying .9966 by the (Pd) of .9 from the model yields .8969. This is not a large reduction from the originally calculated value of .8996 as shown in the model. Again the value of using DOD-HNBK-344/344A for analytical analysis allows calculating the change in screens. Using this understanding along with a spreadsheet program (EXCEL, LOTUS) allows "WHAT IF GAMES" to automatically generate new control charts for that particular screen. The spreadsheet program will be discussed in Section 3 of this document. As was in screen #1 all failures must go through Failure Analysis to correct any major problems during this phase of manufacturing/development (eliminate the hidden factory).

Screen #3 has the same type of input defects. The remaining defects from screen #2 that have escaped detection and other purchased items possibly Government Furnished Equipment (GFE) and interfacing cables. The selection for this screen is 3 thermal cycles (same limits as the previous screens). The same argument stands here as it was in screen #2 about ramp and temperature limits. Ten minutes of random vibration per axis at a reduced Acceleration Spectral Density (ASD) of .02 g2/hz, this yields a vibration profile of 4.3 g's rms. The composite (Pe) is .9663 and the (Pd) is .9 in the model, which yields a screening strength (SS) of .8697. The total

(SS) composite for all three screens combined is .9980. Using this number, it can be said that the Section would contain 0.002 times the total number of defects in the entire system. Failure Analysis and corrective action will be performed on all failures. This is the highest level of assembly in this model and would be a good place to insert Failure Free Acceptance Testing. This would eliminate the need for actual Acceptance Testing, another cost savings.

2.2 OPTIONS

a) Bare board screens – If there are known failure to the bare circuit boards (bad feed thru holes for example) there is no reason not to develop a screen to eliminate this problem before manufacturing the CCA's. This screen would not be cost effective.

b) 100 % parts re-screening – This is not recommended. More problems can be incorporated then eliminated due to Electrostatic Discharge (ESD), damage due to handling and test equipment incompatibility. A screen should be developed to handle known bad components, new vendor components and audit all other parts,

c) Failure Free Acceptance test could be performed at the highest level of assembly to eliminate the need for a separate acceptance test. For example, given the UUT had a yield requirement of .51 or a mean time between failures of 500 hours, the tables of DOD-HNBK-344 would show that the fail free period is 1.9 hours. Therefore the last 2 hours of ESS would be the acceptance test.

John J. Quinn

Cost Effective Analysis is performed using the Cost Analysis Worksheet from DOD-HNBK-344/344A (refer to figure B). This allows all the known costs and expected fallout from all the screens to be tabulated and compared to a threshold cost. If the cost per defect is greater than the threshold cost, the screen is not cost effective.

SECTION 3

C.3 ESS SUMMARY

The Environmental Stress Screen (ESS CCP) Map (refer to Figure 6) is a spreadsheet version of the ESS Model and automatically performs the mathematical operations used to solve the equations required for Precipitation Efficiency, all out, etc. It physically looks like the ESS Model, that is the left hand side of the map represents the CCA/PWA screen. The center section is for the Units/Plate level screen and the right hand side of the map is the Section/Forebody screen. Two additional tasks have been added, they are, 1) The control charts for screen fallout and, 2) Cost analysis to determine if the screen selected is cost effective.

The communications System that was used as an example to develop the spreadsheet used herein is from DOD-HNBK-344/344A. The marriage of the example communications system with the defect density given and a very high precipitation Efficiency (Pe) of the ESS Model may yield unreal values, However it suits the purpose to develop an effective ESS Program with cost requirements and control charts to flag unexpected problems. Some knowledge on generating computer spreadsheets is helpful, this document not intended to prepare/teach basic computer spreadsheets.

John J. Quinn

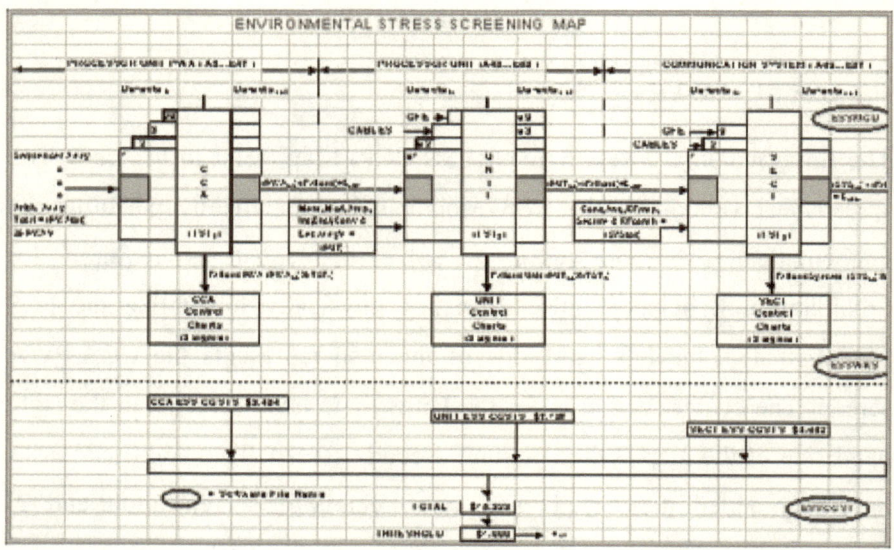

Figure 6 – ESS Screening Map

32

CCP EXPLANATION

The attached three documents: (ESS CCP MAP, ESS MODEL and BREAKDOWN CHART – COMMUNICATION SYSTEM) will be used to develop a C CALC-PLUS computer spreadsheet program.

Looking at the System Breakdown Chart (refer to attachment A-2) the Communications System is divided into three (3) major levels; CCA, Unit and Section. The assembly level is equivalent to a circuit card assembly (CCA).

The Environmental Stress Screen at this level is shown on the ESS Model as screen #1. This screen is 12 Thermal cycles with a range from -45°C to +85°C and a ramp of 15°C/minute. This yields a Precipitation Efficiency (Pe) of 0.9999. The Probability of Detection (Pd) is a function of the effectiveness of the test equipment. For this example the ESS Model shows a (Pd) of 0.85. The Pd is used to calculate Screening Strength (SS).

$$SS = (Pe \ X \ Pd) \qquad \text{Equation \# 1}$$

The latent defects that are detected during the screen is called Fallout. The Fallout (Fo) at this level can be calculated as;

$$Fo = Din \ X \ SS \qquad \text{Equation \# 2}$$

where Din is the total defect density input at this screen.

The defects out (Do) are those defects passed on to the next higher level screen and can be predicted.

$$Do = Din - Fo \qquad \text{Equation \# 3}$$

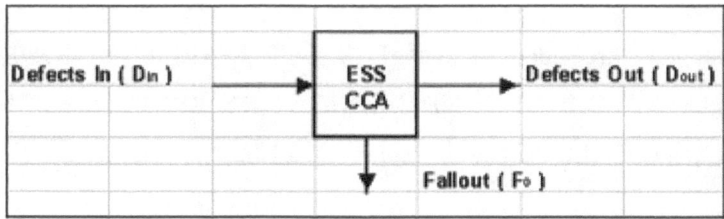

The next higher assembly is shown on the Breakdown Chart as the Processor Unit (refer to attachment A-2). This is shown on the ESS Model as Plate/Unit level, screen #2. The ESS screen at this level has been selected as 6 thermal cycles and 10 minutes of random vibration per axis (3). The random vibration profile is NAVMAT P-9492 which yields 6 g's rms, The thermocycle Precipitation Efficiency is 0.9864 and the vibration Precipitation Efficiency (Pe) is 0.9470 therefore the composite of the two screens is:

$$Pe = 1 - (1 - therm\ Pe)\ (1 - vib\ Pe) \qquad Equation\ \#\ 4$$
$$Pe = 1 - (1 - .9864)\ (1 - .9740)$$
$$Pe = 0.9995$$

a (Pd) at this level has been selected as 0.90, therefore the total Screening Strength (SS) = 0.9995 X 0.9 = 0.8996.

The Processor Unit on the Breakdown Chart shows 9 sub-assemblies that make up the Processor Unit. These sub-assemblies show the part number on the top and the defect density number on the bottom. The Processor Unit shows the total of the nine sub-assemblies defect density as 769,173.5. Using a SS of 0.8996 the Fo and Dout is calculated as:

Fo = Din X SS

Fo = 769,173 X 0.8996

Fo = 691,948 ppm (parts per million)

The Dout passed on to the next higher assembly is:

Dout = Din - Fo

The next higher assembly shown on the Breakout Chart is the Communications System (refer to Attachment A-2). This has 9 units that make up the Communication System. They show part number and defect density number as did the Processor Unit. The Processor Unit is one of the nine units. The total defect density going into the Communication System shows up as 1,414,000 (rounded off). Looking at the ESS Model at Section/Forebody level (screen #3), the ESS screen is 3 thermal cycles and 10 minutes of random vibration per axis (3). The same equation applies as it did at Plate/Unit levels. The Fallout (Fo) calculates to be 0.8697 or 87 % of the latent defects in at this level of screen should be detected. The composite of the Fo at all 3 screens would be 0.998 or 2 possible failures (due to latent defects) per 1,000 systems built would be passed on to the field.

This section will describe how to use a computer spreadsheet, C CALC-P to develop an automated Environmental Stress Screen in accordance with (IAW) DOD-HNBK-344. After the ESS is programmed into the computer it will allow simulation of "What If" scenarios to automatically change the Screening Strength, Fallout, defects passed on to the next level of assembly and their associated control charts. It will allow the user to tailor a screen to a desired result without going through numerous calculations. There are three major spreadsheets involved:

1) ESSMOD – This is sub-divided into 3 work areas that match the 3 levels of Screening from the ESS Model, (1) PWA's, plated wire assemblies/CCA, circuit card assemblies. (2) Unit Level/Plate Level and (3) System/Section Level. These worksheets are described individually in figures 7, 8, 9 and 10. The sequence of these worksheets can then be traced through the system using ESS C CALC-PLUS MAP. The Fallout from each screen is automatically entered into another worksheet named ESSWKS.

2) ESSWKS – This worksheet develops the control charts for each level of screening. The inputs to the control chart are from the ESSMOD file and are automatically entered to develop the +/- 3 sigma limits. The Fallout and Defects Density data are required and a typical screen is shown as follows:

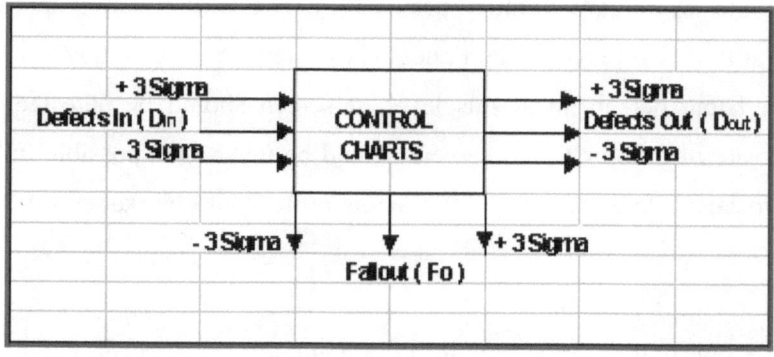

Figure 7 – Defects Out

3) ESS COST – This file will calculate the cost of each level of screening, and then calculate the entire system level costs. This file requires some

manual entries ie, Fixed costs and Variable costs. The bottom line is a comparison of the costs to eliminate defects (cd) vs. The field repair costs (threshold cost – (ct)). If (cd) is greater than (ct) the screen is not cost effective and should be re-examined. If (cd) is less than (ct) the screen is cost effective. For an example refer to Figure 5.

Now refer to the ESS CCP MAP and the printout of file ESSMOD, worksheet (6) region A5.E47 (Figure 8)

Worksheet (6) essmod region: A5...E47				
Processor Unit PWA's//				
A	**B**	**C**	**D**	**E**
Nomenclature	Quantity	Est Defects	Total	Notes
Sequencer assy	1	13,771.80	13,771.80	B5 X C5
Timing assy 1	1	21,256.20	21,256.20	
Timing assy 2	1	19,829	19,829	
Event Sequencr unit	1	13,864.40	13,864.40	
Timing cont assy	1	12,990.20	12,990.20	
Interleave Timing	1	22,791.60	22,791.60	
Interleave assy	1	11,446.80	11,446.80	
Delay assy A	1	26,098.20	26,098.20	
Demod assy	1	60,096.50	60,096.50	
Frequency Cont.	1	5,327.20	5,327.20	
Timing assy 4	1	20,921	20,921	
Quantizer assy	1	24,216.60	24,216.60	
Arithmetic assy	1	2,662.80	2,662.80	B34 X C34
PWA Total ppm		537,285.50	537,285.50	(PWATOT)
Precipitation Efficiency		0.9999	From 344,see model	
Probabilty Detection		0.85	From 344,see model	
Screening Strength		0.8499	> TSC calculation (tst1)	
Fallout ppm		456,638.90	(PWATOT) X (tst1) = (Fallout)	
Defects Out ppm		80,646.50	(PWATOT) - (Fallout) > (Dout1)	
Percent Detected		85%	Fallout (PWATOT)	

Figure 8 – Processor Unit PWA

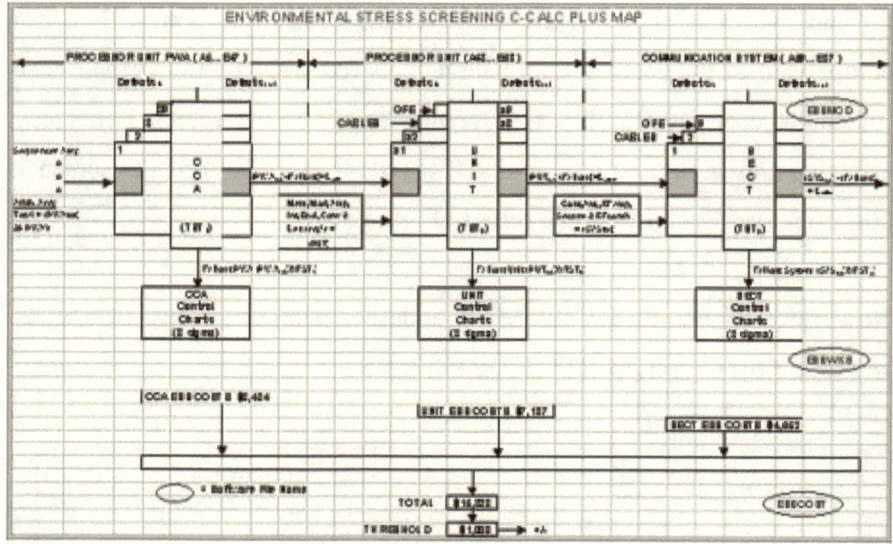

Column A Nomenclature and column B quantity represents the Processor Unit PWA's that make up the Processor Unit. Column C lists the 29 Pwa's defect densities and column D the total defect density in (Din) for the processor Unit. This total is calculated as [sum (d5...d34)] and is labeled PWATOT (printed wire assembly total). In this example the total is 537,785 defects ppm. What is not shown is the defect density for each of the 29 PWA's. The list from DOD-HNBK-344; page 63 was used for this example. Looking at Figure 1, additional calculations can be made.

1) Precipitation Efficiency (Pd) is .9999 from the ESS Model (screen #1).

2) Probability of detection (Pd) of .85 was selected from ESS Model (screen #1).

3)The Screening Strength is calculated as cell C36 X C37 = .8499 in cell C38 and is labeled as (tst1). Labeling allows a name to be used in future calculations. Tst1 representing the calculated SS from screen #1. It is easier to remember a name rather then a cell number.

4) The Fallout is calculated using labels (PWATOT) X (tst1) and is named Fallout.

5) The defects out can be calculated from label (PWATOT − C39, fallout, and is labeled Dout1. This value is the defects passed on to the next higher level of assembly. In this example it will be the Unit Level.

Worksheet (6) essmod Region A48...E68				
Processor Unit				
A	B	C	D	E
Nomenclature	qty	Estimated Defects	Notes	
Processor	1	80,646.50	Input from PWA's	
Comm.	1	74,943.20		
Memory	1	6,855		
Modulator	1	33,738		
Amplifier	1	17,898.50		
Interface	1	2,522.20		
Enclosure	1	8,639.60		
Converter	1	40,405.40		
Detector	1	40,385.50		
Processor Total		306,033.90	(PUT)	
Precipitation Effeciency		0.9995	From 344, see model	
Probability Detection		0.9	From 344, see model	
Screening Strength		0.8996	SS X Pd > TSC Cal (tst2)	
Fallout ppm		275,308.10	(PUT) X (tst2)	
Defects Out		30,725.80	(PUT) - Fallout > (Dout2)	
Percent Detected		90%	Fallout/(PUT)	

Figure 9 – Processor Unit

Column A and Column B (quantity) represent all the components that make up the Processor Unit, the next higher level of assembly from the PWA's. The PWA's that made up the Processor (29 in total) were screened and the defects passed on to the next higher level was named (Dout1), this value is automatically indexed into cell C50 (in this example the value is 80,646 in column C). The remaining inputs to the screen (Communications through Detector) would be handled in the same fashion. Each would have a file that calculates their PWA's defects that pass on to the Processor Unit. They would have been named to automatically enter their values in cells C51 through C58. The Processor Unit total defects are calculated as [sum (C50...C58)] and resides in C59 which has been named Processor Unit Total (PUT). The Screening Strength is calculated from cells C60 X C61 (Pe X Pd, which comes from the ESS Model Plate Level). Screening

Strength (SS) , 8996 is in cell C62 and is named (tst2). Fallout is then calculated from (PUT) X (tst2) and defects out is then (PUT) – Fallout (C63). This is named (Dout2) and is indexed to the next higher level assembly (Communications System).

John J. Quinn

A	B	C	D	E
Worksheet (6) essmod Region: A69..E87				
Communications System				
Nomenclature	**QTY**	**Estimated Defects**	**Notes**	
RF Amp	1	21,983		
Secure Unit	1	8,998.60		
RF Combiner	1	11,197.10		
Control Unit	1	71,154.70		
Antenna Unit	1	86,142.30		
LF Amp Unit	1	225,380.50		
Processor Unit	1	30,725.80	From Processor Unit (Dout2)	
HP Amp Unit	1	155,355.80		
HPA Power Supply	1	64,545.40		
System Total ppm		675,483	(SYSTOT)	
Precipitation Effeciency		0.9663	From 344,see model	
Probability Detecton		0.9	From 344,see model	
Screening Strength		0.8697	SS X Pd > TSC Cal (tst3)	
Fallout ppm		587,467.70	(SYSTOT) X (tst3)	
Defects Out ppm		88,015.40	(SYSTOT) - Fallout (Dout3)	
Percent Detected		87%	Fallout/(SYSTOT)	
A88...E131				
SScomp = 1-(1-SS1)(1-SS2)(1-SS3)				
		1-SS1 = 0.1501		
		1-SS2 = 0.1004		
		1-SS3 = 0.1303		
SS comp = 0.998				
		Communications System		

Figure 10 – Communications System

The description of the Communications System level ESS follows the same process as PWA's (Figure 1) and Units (Figure 2) as previously described. The defects out from the Unit Level ESS are automatically entered in cells C71 through C76. The total is calculated as [sum (C71...C76)] and is in cell C77 and named System Total (SYSTOT). The Precipitation Efficiency and Probability of Detection used are from the ESS Model and are used to calculate Screening Strength [C81 X C82]. The Screening Strength in this example is .8697 and named (tst3). The Fallout

42

is calculated as [(SYSTOT) X (tst3)] and resides in C84. The defects out which is passed to its field life is then [(SYSTOT) – Fallout] and named (Dout3).

The composite of the three screens (pwa's, Units and System) can then be calculated as:

$$Tsc = 1 - (1 - (tst1)) (1 - (tst2)) (1 - (tst3))$$

Tst1, tst2 and tst3 are the Screening Strengths of the 3 screens and named so they are automatically entered into the formula. 1- (tst1) through 1-(tst3) is calculated and the results are multiplied together and subtracted from 1 to yield a composite of .998 or .002 latent defects would reach the field. Remember this is an example, a System from DOD-HNBK-344 with given defect density and a very high Screening Strength from the ESS Model. This may not be true in an actual ESS Program. However, it serves the purpose to develop the spreadsheets.

C.3 CONTROL CHARTS

Procedure E in DOD-HNBK-344 has step by step instructions to estimate the baseline screen control charts. Basically, it recommends 3 control charts for each screen. The determining factors for the 3 sigma control charts are Din (defect density in), the total amount of predicted defects entering into the screen, SS (screening strength) – the Precipitation Efficiency (Pe) X the Detection Efficiency (De). With Din and SS known the calculation for Fallout (Fo) and defects passed on to the next level of assembly (Dout) are as follows:

$$Fo = Din \times SS$$

$$Dout = Din (1 - SS)$$

				Screen # 1			
		D_{In}	▶	SS = .85		▶	D_{out}
	(PWATOT)			(TST1)			
				F_o ▼			

Attachment A was an example of the PWA's (29 of them) and described how to calculate the total defects in (Din) and was labeled (PWATOT). The SS was taken from the ESS Model, screen # 1 and is .85 and labeled (TST1). Now automatically the required data is entered into the control charts (PWATOT) and (TST1) from the file ESS MOD. To calculate the 3 sigma control charts for the first screen the following CCP instructions are shown on figure 3-np.

These calculations would be the control charts for the ESS Model # 1. The calculated Dout along with the other defects listed (purchased assemblies, cables, etc) then become the Din for screen # 2. The same process would then apply for screen # 2 and # 3 of the ESS Model.

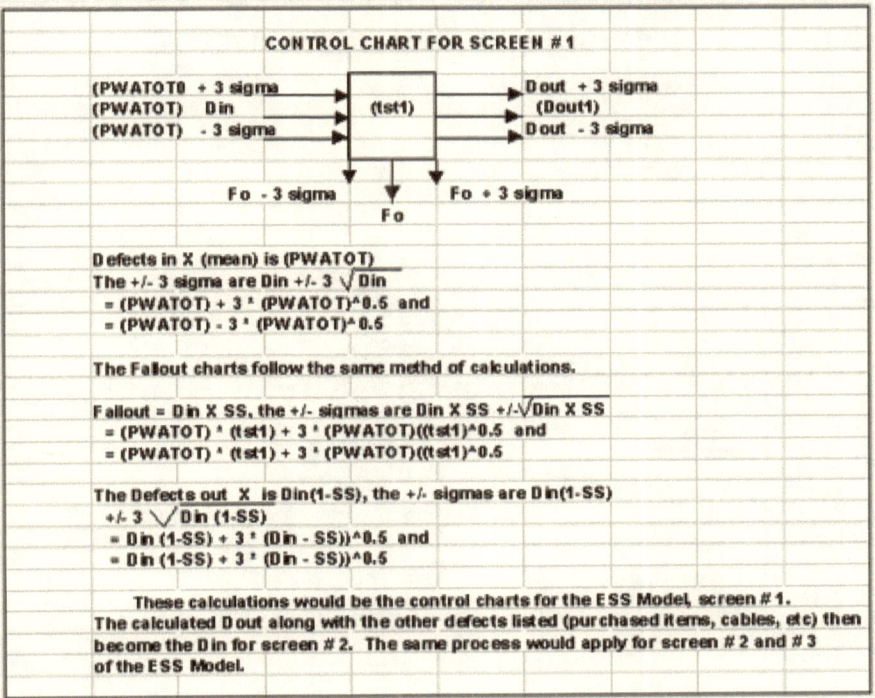

CONTROL CHART FOR SCREEN #1

Defects in X (mean) is (PWATOT)
The +/- 3 sigma are Din +/- 3 $\sqrt{\text{Din}}$
  = (PWATOT) + 3 * (PWATOT)^0.5 and
  = (PWATOT) - 3 * (PWATOT)^0.5

The Fallout charts follow the same methd of calculations.

Fallout = Din X SS, the +/- sigmas are Din X SS +/-$\sqrt{\text{Din X SS}}$
  = (PWATOT) * (tst1) + 3 * (PWATOT)(((tst1)^0.5 and
  = (PWATOT) * (tst1) + 3 * (PWATOT)(((tst1)^0.5

The Defects out X is Din(1-SS), the +/- sigmas are Din(1-SS)
+/- 3 $\sqrt{\text{Din (1-SS)}}$
  = Din (1-SS) + 3 * (Din - SS))^0.5 and
  = Din (1-SS) + 3 * (Din - SS))^0.5

  These calculations would be the control charts for the ESS Model, screen #1.
The calculated Dout along with the other defects listed (purchased items, cables, etc) then
become the Din for screen #2. The same process would apply for screen #2 and #3
of the ESS Model.

Figure 11 – Control Chart Screen #1

45

CONTROL CHART EXAMPLE WORKSHEET

Worksheet: (7) ESSWKS Region: A88..E131

PWA's	PWA's Din		
		Din + 3 sigma =	1.2703
		Din =	0.5372
		Din - 3 sigma =	0
	PWA's Fallout		
		Din X SS + 3 sigma =	2.4841
		Din X SS =	0.4567
		Din X SS - 3 sigma =	-1.5707
	PWA's Dout		
		Din (1-SS) + 3 sigma =	0.9323
		Din(1-SS) =	0.0806
		Din (1-SS) - 3 sigma =	-0.7711
Unit	Unit Din		
		Din + 3 sigma =	1.9655
		Din =	0.306
		Din - 3 sigma =	-1.3535
	Unit Fallout		
		Din X SS + 3 sigma =	1.8494
		Din X SS =	0.2753
		Din X SS - 3 sigma =	-1.2988
	Unit Dout		
		Din (1-SS) + 3 sigma =	0.5563
		Din(1-SS) =	0.0307
		Din (1-SS) - 3 sigma =	-0.4949
System	System Din		
		Din + 3 sigma =	3.1412
		Din =	0.6755
		Din - 3 sigma =	-1.7902
	System Fallout		
		Din X SS + 3 sigma =	2.887
		Din X SS =	0.5875
		Din X SS - 3 sigma =	-1.712
	System Dout		
		Din (1-SS) + 3 sigma =	0.9779
		Din(1-SS) =	0.088
		Din (1-SS) - 3 sigma =	-0.8019

Figure 12 – Control Chart Example

3.3 COST ANALYSIS WORKSHEET

This worksheet is from DOD-HNBK-344 and is combined with the data derived from the Communication System spreadsheets (ESS MOD). The first costs are from the assembly level screen (PWA's column A & C).

1) Fixed and veriable costs are manually entered into cells [c5] & [c6].

2) Expected fallout is automatically entered from ESSMOD cell [d47] divided by 100.

3) The average cost to repair at this level is manually entered, if unknown use $ 40.

4) The screening repair costs are automatically calculated from expected fallout [c7] times cost to repair [c8]. The result is placed into [c9].

5) The assembly level costs are calculated as fixed [c5] plus Variable [c6] plus Repair [c9]. The result is placed in [c10].

The second set of costs are from the Unit Level screens and are as follows:

1) Fixed and veriable costs are manually entered into cells [c13] & [c14].

2) The expected fallout is automatically entered into cell [c15] from file ESSMOD cell [c63] divided by 100.

3) The cost to repair is manually entered into cell [c16] and if unknown use $ 375.

4) The screening repair costs are automatically calculated as [c15] times [c16]. The result is placed in [c17].

5) Unit Level screening costs are then calculated as fixed [c13] plus Variable [c14] plus Repair [c17]. The result is placed into [c18]

The third set of costs are from the System Level screen and are as follows:

1) Fixed and veriable costs are manually entered into cells [c21] & [c22].

2) The expected fallout is automatically entered into cell [c23] from file ESSMOD cell [c87] divided by 100.

3) The cost to repair is manually entered into cell [c24] and if unknown use $ 750.

4) The screening repair costs are automatically calculated as [c23] times [c24]. The result is placed in [c25].

5) System Level screening costs are then calculated as fixed [c21] plus Variable [c22] plus Repair [c25]. The result is placed into [c26].

The Total Screening costs can then be calculated using the following steps:

1) Total Fixed costs = [c5] + [c13] + [c21] = [c29].

2) Total Variable costs = [c6] + [c14] + [c22] = [c30].

3) Total Screening Repair costs = [c9] + [c17] + [c25] = [c31].

4) Total Expected Fallout = [c7] + [c15] + [c23] = [c32].

5) The total number of systems to be screened is manually entered into [c33], in this example it is 100.

6) Total Screening costs = [c10] + [c18] + [c26] = [c34].

7) Screening cost per system = [c34] X 100 = [c35].

8) Cost per defect = [c35] / [c32] = [c36].

9) The threshold cost (average field repair cost) is manually entered into cell [c37], if unknown use $ 1,000.

Now a comparison is made:

If the cost/defect [c36] is greater than the Threshold cost [c37], the screen is not cost effective and the screens should be modified until the cost/defect [c36] is less than the Threshold [c37].

A	B	C	D	E
Worksheet (6) esscost Region: A1..E50				
Cost Analysis Worksheet				
Communications System				
A	**B**	**C**	**D**	**E**
Assembly Screening Costs				
Fixed Screening Cost	*	$1,200		
Variable Screening Cost	*	$2,250		
Expected Fallout @ Assy		0.85		
Avg. Cost/Repair	*	$40		
Unknown use $40				
Screening Repair Cost		$34	(Exp Fo) X Cost/repair	
Assy Level Screening Cost		$3,484	Fixed + Varible + Repair	
Unit Screening Costs				
Fixed Screening Cost	*	$3,750		
Variable Screening Cost	*	$3,100		
Expected Fallout @ Unit		0.9		
Avg. Cost/Repair	*	$373		
Unknown use $375				
Screening Repair Cost		$336	(Exp Fo) X Cost/repair	
Unit Level Screening Cost		$7,186	Fixed + Varible + Repair	
System Screening Costs				
Fixed Screening Cost	*	$1,500		
Variable Screening Cost	*	$2,500		
Expected Fallout @ System		0.9		
Avg. Cost/Repair	*	$750		
Unknown use $750				
Screening Repair Cost		$675	(Exp Fo) X Cost/repair	
System Level Screening Cost		$4,675	Fixed + Varible + Repair	
Total Screening Cost				
Total Fixed Cost		$6,450	Assy + Unit + System	
Total Variable Cost		$7,850	Assy + Unit + System	
Total Repair Cost		$1,045	Assy + Unit + System	
Total Expected Fallout		2.65	Assy + Unit + System	
Number of Systems to Screen	*	100		
Total Screening Cost		$15,345	Fixed + Variable + Repair	
Screening Cost/System		$153	Total Screening Cost/100	
Cost/Defect Eliminated (cd)		$5,790	Total Screening Cost/Tot Fo	
Threshold Cost (ct)	*	$1,000	Related to Field Repair Cost	
If (cd) > (ct) screen is **not** cost effective, If (cd) < (ct) screen **is** cost effective				
* = Manual entry required.				

Figure 13 – Control Chart Example

SYSTEMS BREAKDOWN CHART - COMMUNICATIONS SYSTEM

```
                          3664002
                       Communications
                          System
                       1,415,830.90
```

3664003	3663006	3663013
Processor Unit 769,173.50	Control Unit 71,154.70	RF AMP Unit 21,935.00

3647002	3663025	3511880
Hp Amplifier Unit 155,355.80	Antenna Unit 86,142.50	Secure Unit 8,995.60

3647004	3663040	3663014
HPA Power Supply 64,545.40	LP AMP Unit 225,520.50	RF Combiner Unit 11,157.10

PROCESSOR UNIT BREAKDOWN TO ASSEMBLY LEVEL

```
                          3664005
                        Processor
                          Unit
                       769,173.50
```

3664002	3664003	see table
Enclosure Unit 243.60	Modulator Unit 35,735.00	Processor PWA's 537,726.10

3663004	3664004	
Converter Unit 40,405.40	Amplifier Unit 17,552.50	Comm PWA's 74,945.30

3659007	3663007	3623527
Detector Ass'y 40,535.50	Interface Unit 2,522.20	Memory Unit 9,855.00

Figure 14 – System Breakdown Chart

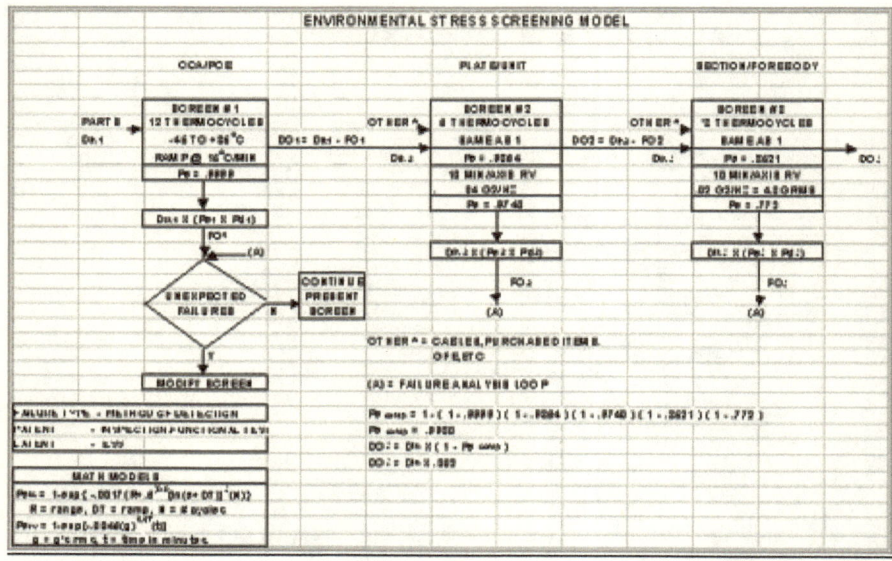

Figure 15 – ESS Model

SECTION 4

4.1 LESSON PLAN/STUDY GUIDE SUMMARY

The information contained within this report will provide the training material for a basic Environmental Stress Screening (ESS) and DOD-HNBK-344 training course. It is presented using the ESS Handbook developed by Thermotron. This handbook should be distributed and read prior to the start of this course. The course intern will start with a review of the handbook to develop baseline ESS knowledge before looking at DOD-HNBK-344. The five major sections of DOD-HNBK-344 will then be discussed.

a) Defect Density and a worksheet will be developed.

b) Determination of Screening Strength using a sample Communications System Block Diagram will be calculated.

c) Failure Free Acceptance Test will be explained using the Phoenix Missile MTBF.

d) Cost Analysis worksheet will be prepared .

e) Monitoring & Control will be discussed to show the on-going task of ESS.

Finally a list of definitions and acronyms from DOD-HNBK-344 is supplied in appendix B.

1) Hold a review of "The Environmental Stress Screening Handbook" provided by Thermotron. This handbook is to be distributed and read prior to training. Review the following topics:

a) The Product Reliability Equation – Life cycle curve.

b) What is ESS? – The screen is not a test – It is a dynamic reliability tool.

c) The Evolution of ESS

 o Laboratory experiments

 o Burn-in

 o MIL-Standards

 o Mission profile testing

 o Thermal cycling

d) Misconceptions about ESS

 o ESS is a test

 o ESS is the same as Burn-in

 o Random sampling

 o Used to validate design

 o ESS induces failures

 o ESS is the same for all products

 o ESS is expensive

e) Types of Environmental Stress Screens

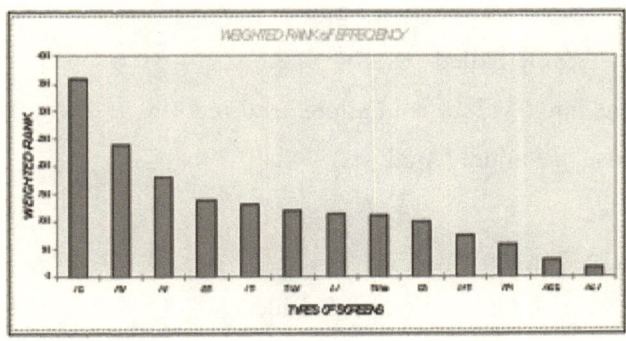

o Temperature cycling

o Random vibration

o Hi temperature Burn-in

o Electrical stress

o Thermal shock

o Sine vibration (fixed frequency)

o Lo temperature soak

o Combined environment

f) Advantages of Temperature Cycling – show charts H & I

g) Levels of complexity

    o Static

    o Dynamic

    o Exercised

    o Full functional

    o Monitored

h) Failure Analysis

John J. Quinn

> o Failure Symptoms
>
> o Environment at Time of Failure
>
> o What Cycle Failed
>
> o Time into Cycle when Failure occurred
>
> o Cause of Failure (Analysis)

I) Tailor ESS to the Products

> o ESS is not the same for all Products

j) Monitor the ESS Program

> o Control Charts

k) Define

> o Aging
>
> o Board level
>
> o Burn-in
>
> o Chamber (thermal)
>
> o Environmental Testing
>
> o Environmental Stress Screening (ESS)
>
> o Failures
>
> o Field Warranty (MBTF)
>
> o Final Assembly level
>
> o Fixturing
>
> o IES
>
> o Latent Defect
>
> o Life Cycle Testing
>
> o Power Cycling (on-off)
>
> o Random Vibration

o RADC

o Screening

o Screening Complexity

o Screening Level

o Screening Strength

o Sine Vibration (Fixed & Swept)

o Stress

o Subassembly level

o Thermal Cycling

l) Using Viewgraph #1 explain the Quantitative Problem and Quantitative Stress Screening Technique

m) Solving the Problem Figure 1-2. (Viewgraph # 2)

      a) Step 1 – Estimate Defects In (Din)

            1) This is procedure A in DOD-HNBK-344

            2) Show Defect Estimation Worksheet (Viewgraph #3)

            3) Defect Tables (parts – MIL-STD-217) (Viewgraph #4)

n) Determine number of defects and yield requirements. This is Procedure B in DOD-HNBK-344.

      a) Show System breakdown chart (Viewgraph #5)

      b) Using (Viewgraph #6) Determine the required Screening Strength (SS) vs. (Viewgraph)

John J. Quinn

5) System.

        o DR = .287

        o SS = .797

        o Pe = .839

c) Show the solution using table (Viewgraph #7) and calculate to verify table.

d) Show example of Composite Screening Strength (Viewgraph #8)

        o Misleading – give example.

e) Explain the Vibration & Thermal defect Detection of (Viewgraph #9)

f) Procedure C – Failure Free Acceptance Test (FFAT)as part of ESS.

g) Using the Viewgraphs go through the calculations to determine FFAT Viewgraphs #10, #11 & #12.

o) Procedure D – using (Viewgraph #13), explain all costs that are considered in an ESS Program.

        a) What are Fixed Costs?

        b) What are Variable Costs?

        c) Threshold Cost = Cost of repair in the field.

p) Procedure E – Monitoring and Control

        a) Explain (Viewgraph #14)

q) Mention the CCALC Plus program being developed

        a) Programs are available for HP handheld calculators.

r) Wrap up discussion using (Viewgraph #15) – bottom line savings in K $.

        a) CCA level - -17.5

        b) Plate level – 14.1

c) Section level – 182.0

d) Total savings – 178.6

John J. Quinn

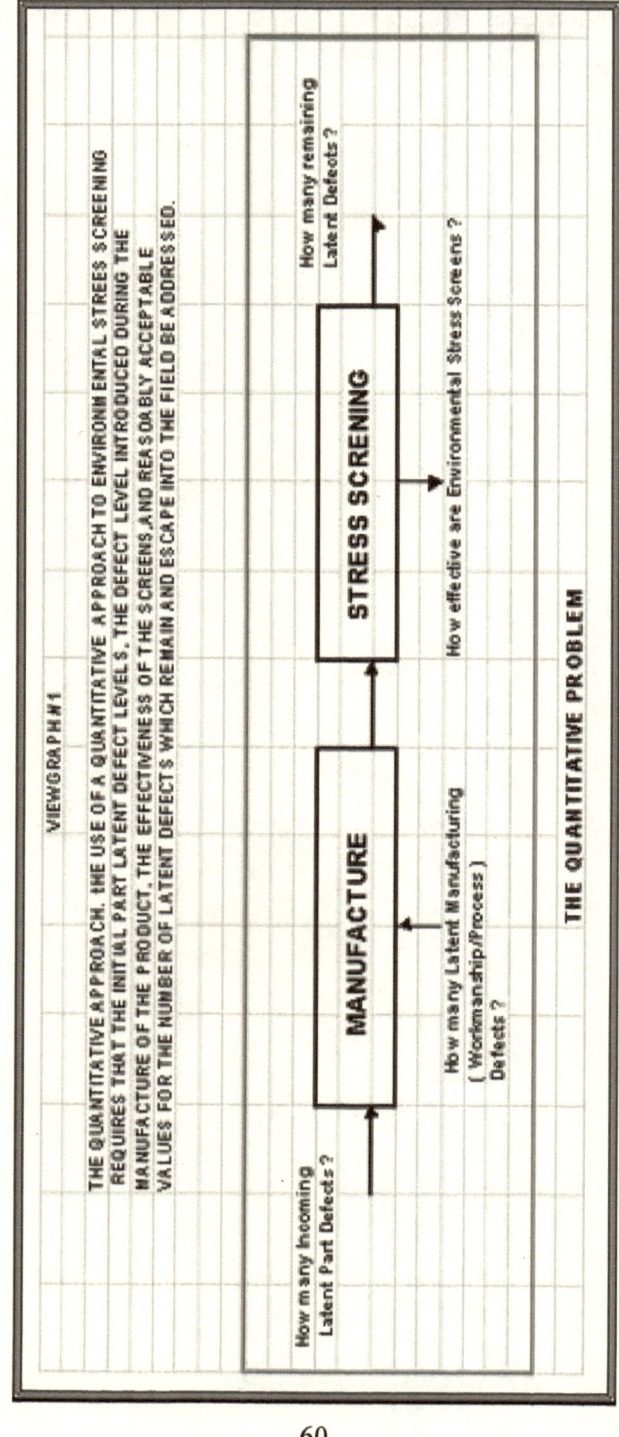

VIEWGRAPH #1

THE QUANTITATIVE APPROACH. THE USE OF A QUANTITATIVE APPROACH TO ENVIRONMENTAL STRESS SCREENING REQUIRES THAT THE INITIAL PART LATENT DEFECT LEVELS, THE DEFECT LEVEL INTRODUCED DURING THE MANUFACTURE OF THE PRODUCT, THE EFFECTIVENESS OF THE SCREENS, AND REASONABLY ACCEPTABLE VALUES FOR THE NUMBER OF LATENT DEFECTS WHICH REMAIN AND ESCAPE INTO THE FIELD BE ADDRESSED.

How many Incoming Latent Part Defects?

MANUFACTURE

How many Latent Manufacturing (Workmanship/Process) Defects ?

STRESS SCREENING

How effective are Environmental Stress Screens ?

How many remaining Latent Defects ?

THE QUANTITATIVE PROBLEM

60

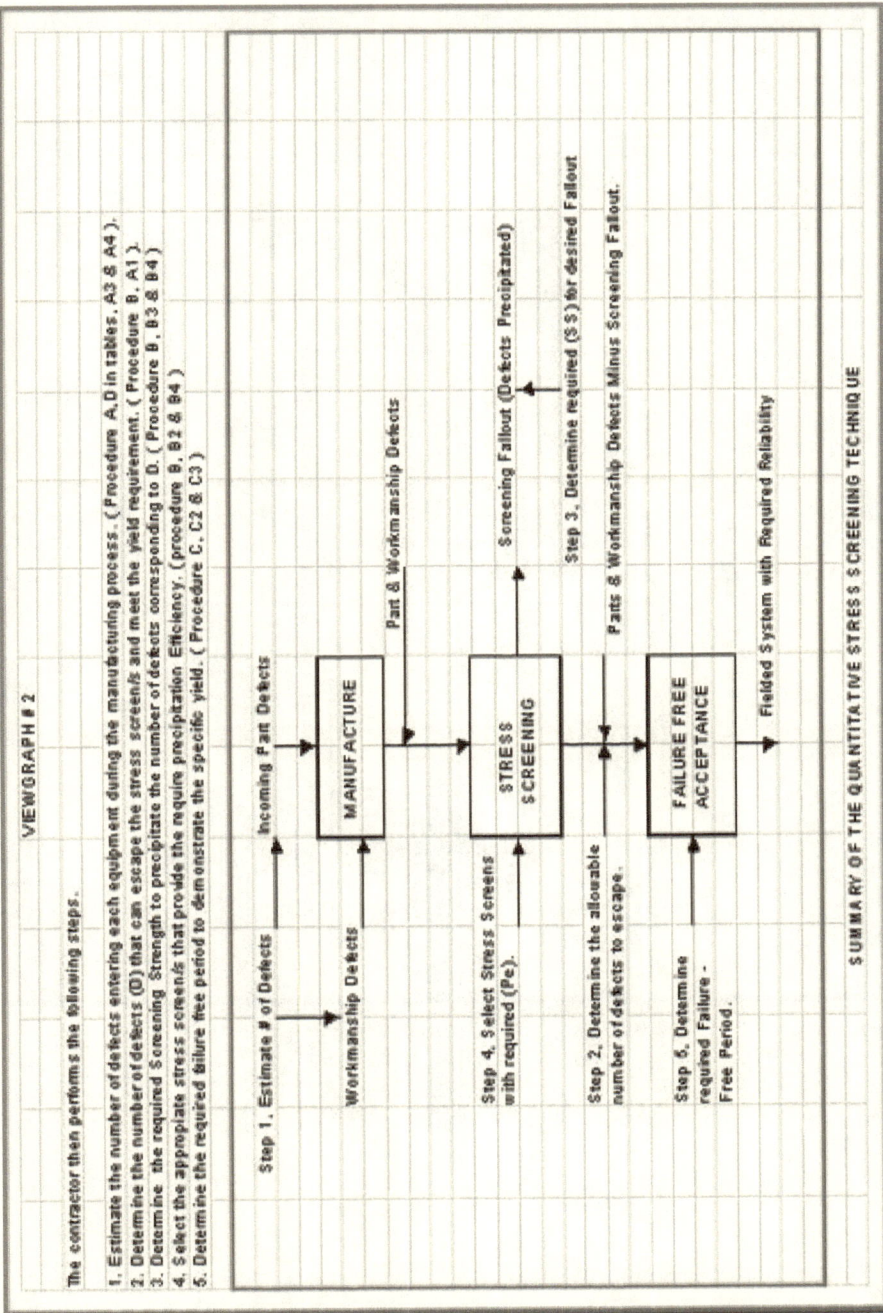

VIEWGRAPH # 2

The contractor then performs the following steps.

1. Estimate the number of defects entering each equipment during the manufacturing process. (Procedure A, D in tables. A3 & A4).
2. Determine the number of defects (D) that can escape the stress screens and meet the yield requirement. (Procedure B, A1).
3. Determine the required Screening Strength to precipitate the number of defects corresponding to D. (Procedure B, B3 & B4)
4. Select the appropiate stress screens that provide the require precipitation Efficiency. (procedure B, B2 & B4)
5. Determine the required failure free period to demonstrate the specific yield. (Procedure C, C2 & C3)

Incoming Part Defects

Step 1. Estimate # of Defects

Workmanship Defects

MANUFACTURE

Part & Workmanship Defects

STRESS
SCREENING

Screening Fallout (Defects Precipitated)

Step 3. Determine required (S S) for desired Fallout

Parts & Workmanship Defects Minus Screening Fallout.

Step 4. Select Stress Screens with required (Pe).

Step 2. Determine the allowable number of defects to escape.

FAILURE FREE
ACCEPTANCE

Step 5. Determine required Failure - Free Period.

Fielded System with Required Reliability

SUMMARY OF THE QUANTITATIVE STRESS SCREENING TECHNIQUE

61

DEFECT ESTIMATION WORKSHEET

Unit Processor unit	Assembly Interface	Prepared by	Date 3/28/96

Part Type	Quality Level	Quantity	Fraction Defective	Estimated Defects*
Microelectronic	B - o	49	87.0	4263.0
Transistors				
Diodes	JANTX	1	46.9	46.9
Resistors	ER - M	18	23.8	428.4
Capacitors	ER - M	1	115.3	115.3
Inductive Devices				
Rotating Devices				
Relays				
Switches				
Connectors	M/S	1	168.0	168.0
Printed Wiring Boards	M/S	1	1139.3	1139.3

FIGURE 1-1 SAMPLE WORKSHEET

JJ Quinn

PART FRACTION DEFECTIVE , CAPACITORS (ppm)

ENVIRONMENT	QUALITY LEVEL					MIL-SPEC	LOWER
	S	R	P	M	L		
GB	1.2	3.8	11.5	38.4	115.3	115.3	384.4
GF	1.8	6.2	18.4	61.5	194.5	184.5	615
GM	9	30	89.9	299.8	899.4	899.4	2998.1
MP	12.7	42.3	126.8	422.9	1268.4	1268.4	4228.1
MSB	5.8	19.2	57.7	192.2	576.6	576.6	1926.9
MS	6.3	21.1	63.4	211.4	634.2	634.2	2114.1
MU	14.3	47.7	143	476.4	1429.9	1429.9	4766.2
MH	18.4	61.5	184.5	615	1845	1845	6150
MUU	20.8	69.2	207.6	691.9	2075.6	2075.6	6918
ARW	27.7	92.2	276.7	922.5	2767.5	2767.5	9225
AIB	5.8	19.2	57.7	192.2	576.6	576.6	1921.9
AIA	3.5	11.5	34.6	115.4	345.9	345.9	1152.1
SF	.9	3.1	9.2	30.1	92.2	92.2	307.5

VIEWGRAPH # 4

SYSTEMS BREAKDOWN CHART - COMMUNICATIONS SYSTEM

9664008
Communications System
1,413,930.90

9664009 Processor Unit 769,173.50	9663006 Control Unit 71,154.70	9663013 RF AMP Unit 21,983.00
9647002 Hp Amplifier Unit 155,355.80	9663025 Antenna Unit 86,142.30	9511880 Secure Unit 8,998.60
9647004 HPA Power Supply 64,545.40	9663040 LP AMP Unit 225,380.50	9663014 RF Combiner Unit 11,197.10

PROCESSOR UNIT BREAKDOWN TO ASSEMBLY LEVEL

9664009
Processor Unit
769,173.50

9664002 Enclosure Unit 263.60	9664003 Modulator Unit 33,738.00	see table Processor PWA's 537,786.10
9663004 Converter Unit 40,405.40	9664004 Amplifier Unit 17,898.50	Comm PWA's 74,043.20
9650067 Detector Assy 40,335.50	9663007 Interface Unit 2,522.20	9629327 Memory Unit 5,855.00

PROCEDURE B

DETERMINE REQUIRED SCREENING STRENGTH (SS) GIVEN YIELD = 0.75
SYSTEM LEVEL CONTAINS 1.414 DEFECTS (Din).

$$Dout = - \ln (Yield)$$
$$Dout = - \ln (0.75)$$
$$Dout = 0.287$$

A SYSTEM WHICH HAS 1.414 DEFECTS SHOULD BE SCREENED SO THAT AN
AVERAGE OF ONLY 0.287 DEFECTS REMAIN.

$$Dout = Din (1 - SS)$$
$$0.287 = (1 - SS)$$
$$SS = 1 - (0.287/1.414)$$
$$SS = 0.797$$

SS = Pe X De, De IS DETECTION EFFECIENCY = .95

$$Pe = SS/De$$
$$Pe = 0.797/.95$$
$$Pe = .839$$

USE THE TEMPERATURE TABLE 2.20 WITH A RANGE OF 120°C, RAMP OF 20°C
AND THE REQUIRED NUMBER OF CYCLES = 2.

$$Pe = .8407$$

VIEWGRAPH # 6

65

PRECIPITATION EFFECIENCY - THERMAL CYCLING

TEMPERATURE RANGE (°C)

Number of Cycles	Ramp oC per minute	20	40	60	80	100	120	140	160	180
2	5	.1633	.2349	.2886	.3324	.3697	.4023	.4312	.4572	.4809
	10	.2907	.4031	.4812	.5410	.5891	.6290	.6629	.6920	.7173
	15	.3911	.5254	.6124	.6752	.7232	.7612	.7920	.8175	.8388
	20	.4707	..6155	.7034	.7636	.8075	.8407	.8665	.8871	.9037
4	5	.2998	.4141	.4939	.5543	.6027	.6427	.6765	.7054	.7305
	10	.4969	.6437	.7308	.7893	.8312	.8624	.8863	.9051	.9201
	15	.6292	.7748	.8498	.8945	.9234	.9430	.9567	.9667	.9740
	20	.7198	.8522	.9120	.9441	.9620	.9746	.9822	.9875	.9907
6	5	.4141	.5522	.6400	.7025	.7496	.7864	.8160	.8401	.8601
	10	.6431	.7873	.8603	.9033	.9306	.9489	.9617	.9708	.9774
	15	.7742	.8931	.9418	.9657	.9788	.9864	.9910	.9939	.9958
	20	.8517	.9432	.9739	.9868	.9929	.9960	.9976	.9986	.9991
8	5	.5098	.6574	.7439	.8014	.8422	.8723	.8953	.9132	.9274
	10	.7469	.8731	.9275	.9556	.9715	.9811	.9871	.9910	.9936
	15	.8625	.9493	.9774	.9889	.9941	.9967	.9981	.9989	.9993
	20	.9215	.9781	.9923	.9969	.9986	.9994	.9997	.9998	.9999
10	5	.5898	.7379	.8179	.8674	.9005	.9237	.9405	.9529	.9623
	10	.8204	.9242	.9624	.9796	.9883	.9930	.9956	.9972	.9982
	15	.9163	.9759	.9913	.9964	.9904	.9992	.9996	.9988	.9999
	20	.9585	.9916	.9977	.9993	.9997	.9999	.9999	.9999	.9999
12	5	.6568	.7994	.8704	.9115	.9373	.9544	.9661	.9744	.9804
	10	.8725	.9548	.9805	.9906	.9952	.9974	.9985	.9991	.9995
	15	9490	.9886	.9966	9988	.9996	.9998	.9999	.9999	.9999
	20	.9780	.9968	.9993	.9998	.9999	.9999	.9999	.9999	.9999

Ng Values for Temperature Cycling
Temperature Range

Rate of change	20	40	60	80	100	120	140	160	180
5	.0891	.1339	.1703	.202	.2308	.2573	.2821	.3055	.3278
10	.1717	.258	.3281	.3893	.4447	.4958	.5436	.5888	.6317
15	.248	.3726	.4739	.5623	.6423	.7161	.7852	.8504	.9125
20	.3181	.4779	.6077	.7212	.8237	.9184	1.007	1.0906	1.1702

VIEWGRAPH # 7

VIEWGRAGH # 8

COMPOSITE SCREENING STRENGTH

o EXAMPLE - LET AN ESS PROGRAM BE DISTRIBUTED TO THREE LEVELS (MODULES, UNITS AND SYSTEM) AND EACH HAS A CONSERVATIVE SCREENING STRENGTH OF .60, THEN;

$$SS_{comp} = 1 - (1 - SS_1) (1 - SS_2) (1 - SS_3)$$

$$SS_{comp} = 1 - (1 - .6) (1 - .6) (1 - .6)$$

$$SS_{comp} = 1 - .064 = .969$$

ASSEMBLY DEFECT TYPES PRECIPITATED BY THERMAL
AND VIBRATION SCREENS

FIGURE 1-2 Reference RADC TR 82-87

Defect Type	Thermal Screen	Vibration Screen
Defective Part	X	X
Broken Part	X	X
Improperly Installed Part	X	X
Solder Connection	X	X
PCB Etch - shorts & opens	X	X
Loose Contact		X
Wire Insulation	X	
Loose Wire Termination	X	X
Improper Crimp	X	
Contamination	X	
Debris		X
Loose Hardware		X
Chafed /pinched wires		X
Parameter Drift	X	
Hermetic Seal Failure	X	
Adjacent Boards/Parts Shorting		X

VIEWGRAPH # 9

68

**ASSEMBLY DEFECT TYPES PRECIPITATED BY THERMAL
AND VIBRATION SCREENS**

FIGURE 1-2 Reference RADC TR 82-87

Defect Type	Thermal Screen	Vibration Screen
Defective Part	X	X
Broken Part	X	X
Improperly Installed Part	X	X
Solder Connection	X	X
PCB Etch - shorts & opens	X	X
Loose Contact		X
Wire Insulation	X	
Loose Wire Termination	X	X
Improper Crimp	X	
Contamination	X	
Debris		X
Loose Hardware		X
Chafed /pinched wires		X
Parameter Drift	X	
Hermetic Seal Failure	X	
Adjacent Boards/Parts Shorting		X

VIEWGRAPH # 9

VIEWGRAPH # 10

PROCEDURE C

o FAILURE FREE TEST PERIOD (PHEONIX MISSILE MBTF)

- MBTF 500 HRS (GIVEN)

- YIELD = .51 (TABLE 2.27)

o FAILURE FREE PERIOD = Ng

Ng = 1/MBTF = 1/500 = .002

Tf = FAILURE RATE RATIO (FROM Pe TABLE) - .7852

FAILURE RATE RATIO = Tf/Ng = .7852/.002 = 392.6

USING TABLE 2.26 FOR FAIL TIME (Ft) = 1.5 HOURS

FAILURE FREE TIME = 1.5/.7852 = 1.91 HOURS

THIS IS FOR A YIELD OF .51 WITH A 90 % LOWER CONFIDENCE BOUND. THE PRODUCT MUST OPERATE 2 HOURS FAILURE FREE DURING THE STRESS SCREEN.

70

PROCEDURE C

o FAILURE FREE TEST PERIOD (PHEONIX MISSILE MBTF)

    - MBTF 500 HRS (GIVEN)

    - YIELD = .51 (TABLE 2.27)

o FAILURE FREE PERIOD = Ng

    $Ng = 1/MBTF = 1/500 = .002$

    Tf = FAILURE RATE RATIO (FROM Pe TABLE) - .7852

    FAILURE RATE RATIO = Tf/Ng = .7852/.002 = 392.6

    USING TABLE 2.26 FOR FAIL TIME (Ft) = 1.5 HOURS

    FAILURE FREE TIME = 1.5/.7852 = 1.91 HOURS

    THIS IS FOR A YIELD OF .51 WITH A 90 % LOWER CONFIDENCE BOUND. THE PRODUCT MUST OPERATE 2 HOURS FAILURE FREE DURING THE STRESS SCREEN.

	90 % LOWER CONFIDENCE BOUND ON YIELD (1-60)												
	FAILURE RATE RATIO												
Tr	1	2	3	4	5	6	7	8	9	10	20	40	>60
1	0.47	0.35	0.32	0.3	0.29	0.29	0.28	0.28	0.28	0.28	0.27	0.27	0.26
1.1	0.55	0.42	0.36	0.35	0.35	0.35	0.34	0.34	0.34	0.33	0.33	0.32	0.32
1.2	0.62	0.48	0.44	0.42	0.41	0.4	0.4	0.4	0.39	0.39	0.39	0.38	0.37
1.3	0.69	0.54	0.5	0.48	0.47	0.46	0.45	0.45	0.44	0.44	0.43	0.43	0.43
1.4	0.74	0.59	0.55	0.53	0.52	0.51	0.5	0.5	0.5	0.49	0.48	0.48	0.47
1.5	0.72	0.64	0.6	0.57	0.56	0.55	0.55	0.54	0.54	0.54	0.53	0.52	0.52
1.6	0.84	0.66	0.64	0.62	0.61	0.6	0.59	0.59	0.58	0.58	0.57	0.56	0.56
1.7	0.87	0.72	0.66	0.66	0.64	0.64	0.63	0.63	0.62	0.62	0.61	0.6	0.6
1.8	0.91	0.79	0.75	0.73	0.71	0.71	0.7	0.7	0.69	0.69	0.68	0.67	0.67
1.9	0.93	0.79	0.75	0.73	0.71	0.71	0.7	0.7	0.69	0.69	0.68	0.67	0.67
2	0.95	0.82	0.77	0.75	0.74	0.73	0.73	0.73	0.72	0.72	0.71	0.7	0.7
2.2	0.99	0.86	0.82	0.8	0.79	0.79	0.78	0.78	0.77	0.77	0.76	0.76	0.75
2.4	1	0.9	0.85	0.84	0.83	0.83	0.82	0.82	0.82	0.81	0.8	0.8	0.8
2.6	1	0.92	0.89	0.88	0.87	0.85	0.85	0.85	0.85	0.85	0.84	0.84	0.83
2.8	1	0.94	0.92	0.9	0.89	0.88	0.88	0.88	0.88	0.87	0.87	0.87	85
3	1	0.96	0.93	0.92	0.91	0.91	0.91	0.9	0.9	0.9	0.89	0.89	0.89
3.5	1	0.98	0.97	0.96	0.95	0.95	0.95	0.94	0.94	0.94	0.94	0.93	0.93
4	1	0.99	0.93	0.98	0.97	0.97	0.97	0.97	0.97	0.97	0.96	0.96	0.96
5	1	1	1	1	0.99	0.99	0.99	0.99	0.99	0.99	0.99	0.99	0.99
6	1	1	1	1	1	1	1	1	1	1	1	0.99	0.99
7	1	1	1	1	1	1	1	1	1	1	1	1	1

VIEWGRAPH # 12

72

COST WORKSHEET

SYSTEM/PROJECT
Prepared by: **Date**
ASSEMBLY SCREENING COSTS

1. Fixed Screening Cost	$_____
2. Variable Screening Cost	$_____
3. Expected Assembly Level Fallout	_____
4. Average Cost per Repair (If unknown use $40)	$_____
5. Screening Repair Cost (multiply line 3 by line 4)	$_____
6. Assembly Level Screening Cost (add lines 1,2 & 5)	$_____

UNIT SCREENING COSTS

1. Fixed Screening Cost	$_____
2. Variable Screening Cost	$_____
3. Expected Unit Level Fallout	_____
4. Average Cost per Repair (If unknown use $375)	$_____
5. Screening Repair Cost (multiply line 3 by line 4)	$_____
6. Unit Level Screening Cost (add lines 1,2 & 5)	$_____

SYSTEM SCREENING COSTS

1. Fixed Screening Cost	$_____
2. Variable Screening Cost	$_____
3. Expected System Level Fallout	_____
4. Average Cost per Repair (If unknown use $750)	$_____
5. Screening Repair Cost (multiply line 3 by line 4)	$_____
6. System Level Screening Cost (add lines 1,2 & 5)	$_____

TOTAL SCREENING COST

7. Total Fixed Cost	$_____
8. Total Variable Cost	$_____
9. Total Screening Repair Cost	$_____
10. Total Expected Fallout	_____
11. Total Number of Systems to be Produced	_____
12. Total Screening Cost (add lines 7,8 & 9)	$_____
13. Total Screening Cost per System (divide line 12 by 11)	$_____

COST PER DEFECT ELIMINATED (divide line 12 by 10) $_____

THRESHOLD COST (average cost to repair in field) $_____

If the cost per defect is greater than the threshold cost, $C_d > C_t$ the screen is <u>not</u> cost effective and should be re-evaluated.

VIEWGRAPH # 13

73

SUGGESTIONS FOR REVISING THE EXPECTED DEFECT PRECIPITATION ESTIMATES BASED ON OBSERVED RESULTS								
Location of Observed Defects Relative to Probability Interval			Alternative Recommended Actions					
			Assembly		Unit		System	
Assembly	unit	System	ID	SS	ID	SS	ID	ss
Within	Within	Within	NC	NC	NC	NC	NC	NC
Within	Within	Above	NC	NC	NC	NC	I	NC
Within	Within	Below	NC	NC	NC	NC	D	NC
Within	Above	Within	NC	NC	I	NC	NC	NC
Within	Above	Above	NC	NC	I	NC	NC	NC
Within	Above	Below	NC	NC	I	NC	NC	NC
Within	Below	Within	NC	NC	D	NC	NC	NC
Within	Below	Above	NC	NC	NC	D	NC	NC
Within	Below	Below	NC	NC	D	NC	NC	NC
Above	Within	Within	NC	I	NC	NC	NC	NC
Above	Within	Above	NC	I	NC	NC	I	NC
Above	Within	Below	NC	I	NC	NC	D	NC
Above	Above	Within	I	NC	NC	NC	NC	NC
Above	Above	Above	I	NC	I	NC	NC	NC
Above	Above	Below	I	NC	NC	I	NC	NC
Above	Below	Within	NC	I	NC	NC	NC	NC
Above	Below	Above	I	I	NC	D	NC	NC
Above	Below	Below	D	NC	NC	NC	NC	NC
Below	Within	Within	NC	D	NC	NC	NC	NC
Below	Within	Above	D	NC	NC	NC	D	NC
Below	Within	Below	NC	D	NC	NC	NC	NC
Below	Above	Within	NC	D	NC	NC	NC	NC
Below	Above	Above	NC	D	NC	NC	I	NC
Below	Above	Below	D	NC	NC	NC	NC	NC
Below	Below	Within	NC	D	NC	NC	NC	NC
Below	Below	Above	D	NC	D	NC	NC	NC
Below	Below	Below	NC	NC	NC	NC	NC	NC
NC = No Change I = Increase Est, D = Decrease Est. ID = Incoming Defects								
SS = Screenin Strength								
Viewgraph # 14								

74

Manufacturing Stress Screening Cost Analysis Example

	Assembly Level		Unit Level		System Level	
	W/O SS	W/SS	W/O SS	W/SS	W/O SS	W/SS
Number of Defects Precipitated	140	490	264	217	268	86
Cost per Repair	$50	$50	$300	$300	$1,000	$1,000
Repair Cost (K$)	7	24.5	79.2	65.1	268	86

Viewgraph # 15

John J. Quinn

SECTION 5

C.3 ENVIRONMENTAL STRESS SCREENING EQUATIONS

This section will describe how to use the EXCEL program "ESSEQ.XLS". The Thermal Cycling algorithm is located from A4 through I9 (refer to Figure 16). The Constant Temperature (Soak) algorithm is located from A11 through I15. The Random Vibration algorithm is located from A17 through I23, The Swept Sine algorithm is found at A26 through I31, and finally the Single Frequency Vibration algorithm is from A34 through I39. An example of Thermal Cycling will be shown:

		SCREENING STRENGTH EQUATONS			
		For Thermal Cycling, 1-exp{-.0017(R+.8)$^{.6}$[ln(e+D)]^3(N)}, enter the temperature range in degrees C in b6, the number of cycles in b7 and the ramp in degrees C/minute in b8. If the unit is operating, enter 1.0 in b9. If non-operating, enter 0.5.			
range	120				
cycles	5	Screening Strength =	0.9160		
ramp	10	Total.. =	0.2864		
op/non-op	0.5				
		For Constant Temperature (soak), 1-exp[-.0017(R+.8)$^{.6}$(T)] enter the temperature range in degrees C in b12 and the time in hours in b13. If the unit is operating, enter 1.0 in b14. If non-operating enter 0.5.			
range	150				
time	30	Screening Strength =	0.9021		
op/non-op	0.5	Total.. =	0.9208		
		For Random Vibration, 1-exp[-.0046(G)$^{1.71}$(T)] enter the g's rms in b21 and the time in minutes in b22. If the unit is operating, enter 1.0 in b23. If non-operating enter 0.5.			
g rms	12				
time	30.0	Screening Strength =	0.9999		
op/non-op	1	Total.. =	0.2000		
		For Swept Sine Vibration, 1-exp[-.000727(G)$^{.863}$(T)] enter the g's in b29 and the time in minutes in b30. If the unit is operating, enter 1.0 in b31. If non-operating enter 0.5.			
g rms	15				
time	30.0	Screening Strength =	0.1972		
op/non-op	1	Total.. =	0.0896		
		For Single Frequency Vibration, 1-exp[-.00047(G)$^{.49}$(T)] enter the g's in b36, time in minutes in b37. If the unit is operating, enter 1.0 in b38. If non-operating enter 0.5.			
g rms	12				
time	30.0	Screening Strength =	0.0466		
op/non-op	1	Total.. =	0.0098		
		SCREENING STRENGTH EQUATONS (con't)			

		COMPOSITES = 1-(1-ss$_i$)...(1-ss$_i$)		
	Place the screening strengths in column D in the composite equation selected			
	ss # 1	0.366+	Composite = 0.4931	
	ss # 2	0.2		
	ss # 1	0.234+5	Composite = 0.6609	
	ss # 2	0.3+56		
	ss # 3	0.123+		
	ss # 1	0.234+5	Composite = 0.7680	
	ss # 2	0.3+56		
	ss # 3	0.123+		
	ss # 4	0.4+++		

Figure 16 – ESSEQ.xls

A	B	C	D	E	F	G	H	I
			For Thermal Cycling, $1-\exp\{-.0017(R+.6)^6[\ln(e+D)]^3(N)\}$, enter the temperature range in degrees C in b6, the number of cycles in b7 and the ramp in degrees C/minute in b8. If the unit is operating, enter 1.0 in b9. If non-operating, enter 0.5					I4 & I5
range	120							I6
cycles	5		Screening Strength =		0.9160			I7
ramp	10		Total$_{ss}$ =		0.3664			I8
op/non-op	0.5							I9

Figure 17 – Thermal Cycling Algorithm

In this EXAMPLE (Figure 17) the temperature range is specified as 120°C and is entered in cell B6. The number of temperature cycles 1s five and is entered in cell B7 and the Ramp (change in temperature per minute) is 10 (10°C/minute). The algorithm is then automatically calculated and the results are shown in cell F7, in this case 0.9160. If the screen is operating, that is electrical testing is performed during the environmental screen, then 1.0 would have been entered in cell B9. This example shows this screen is non operating and the best case is 50 % detection. This is a temperature cycling screen and will detect 80 % of the total "latent defects". Therefore, the total Screening Strength is 0.9160 X 0.5 X 0.8 = 0.3664. This again is an automatic calculation and is found in cell F8.

An example of Random Vibration will now be shown in Figure 18. The Random Vibration algorithm requires as an input G's rms (the square root of the area under the vibration curve) and is entered in cell B21. The other variable is vibration time in minutes and is entered in cell B22. The Screening Strength is automatically calculated as 0.9999 and is located in cell F22. This example is an operating screen, therefore 1.0 is entered in cell B23. Random Vibration will detect 20 % of all "latent defects". The total SS is then 0.9999 X 1.0 X 0.2 = 0.2000 and is found in cell F23.

A	B	C	D	E	F	G	H	I
			For Random Vibration, 1-exp[-.0046(G)$^{1.71}$(T)] enter the g's rms in b21 and the					I18
			time in minutes in b22. If non-operating enter 0.5.					
			the time in minutes in b22. If the unit is operating, enter 1.0 in b2					I19
			If non-operating enter 0.5.3.					I20
g rms	12							I21
time	30.0		Screening Strength =	0.9999				I22
op/non-op	1		Total$_{ss}$ =	0.2000				I23

Figure 18 – Random Vibration Algorithm

The next topic is Composites (figure 19), this calculation is the combination of screening strengths of two or more stress screens. This example has a thermal cycle SS of.03664 (entered in cell D50) and a random vibration SS of 0.2000 (entered in cell D51). The composite of the two screens is automatically calculated as 0.4931 and is in cell G50.

```
B      C      D      E       F        G               H
              COMPOSITES = 1-(1-ss₁)...(1-ssₙ)

   Place the screening strengths in column D in the composite equation
selected

B50 ss # 1     0.3664      Composite =   0.4931
B51 ss # 2       0.2
```

Figure 19 – Composite Algorithm

The next topic is Damage Index (figure 20), how much life was used in ESS. This example will have a life cycle of 500 thermal cycles (C75) and a temperature range of 100°C (C76). The ESS thermal screen is 5 cycles (G75) and a temperature range of 120°C (G76). This example calculates ess/life to be 3 %. That is the thermal ESS has used 3 % of the units life.

```
B          C          D          E          F        G        H
              DAMAGE INDEX (thermal ; d=ns^2.5)
                   n= # cycles &  s= range
        life                                      ess
#Cycles    500.00                        # Cycles   5.00    H75
Range °C   100.00                        Range °C   120.00  H76
                   ess/life = 3%                            H77
```

Figure 20 – Thermal Damage Index

An example of Vibration Damage Index (Figure 21) has the life time (minutes) entered into cell (C82) and life grms entered in cell (C83). The ESS values of time is in cell (G82) and grms in cell (G83). The Damage

Index is calculated as 15 % and is in (E84). This vibration screen uses 15 % of the units life cycle.

B	C	D	E	F	G	H
		DAMAGE INDEX (vibration ; $d=ns^{6.4}$)				
		n= # time & s= grms				
Life				ess		
time/min	120			time/min	30	H82
g's rms	6.5			g's rms	6	H83
		ess/life = 15%				H84

Figure 21 – Vibration Damage Index

Now to plot the screening Strength of this example (two variables). Use an EXCEL spreadsheet and list the screening strengths in column A, place the thermal and vibration data in columns B and C. Click on the chart wizard and follow instructions to plot the graph as shown in Figure

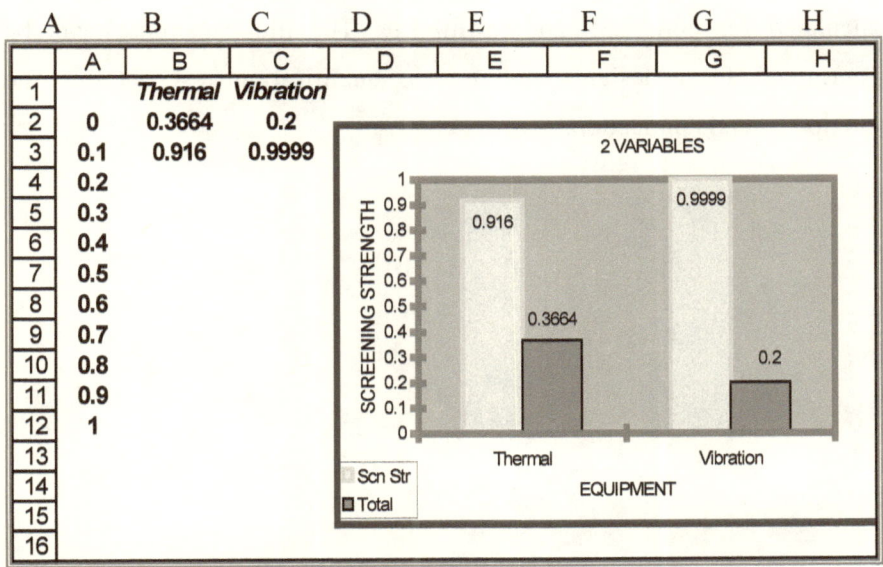

Figure 22 – EXCEL Graph

A brief summary of the graph, column A contains the Yaxis of the bar graph (screening strength). Columns B and C contain the title of the screen, the screening strength calculations and the total screening strength after the op/non-operating and screen type factors are plugged in the calculations (refer to thermal cycling and vibration examples on page). Change any variable in columns B and C and the program will automatically generate a new bar graph.

5.2 Damage Index

After optimizing the screen selection and placement, it is necessary to ensure that the ESS is not too stressful and does not consume too much of the useful (fatigue) life. This is determined by calculating the damage index D from the equation $D=NS^b$, where N represents the stress duration, S = Stress level and b = the fatique exponent. The damage Index should be calculated for both ESS and useful life. The life capabilities can be determined from design requirements, qualification test data or the anticipated end application.

B	C	D	E	F	G
		DAMAGE INDEX (thermal ; $(d=ns^{2.5})$			

	Life			ess	
# Cycles (n)	7300		# Cycles (n)	5.00	
Range (s)	30		Range (s)	120.00	

$$ess/life = 21.9\ \%$$

$$\text{DAMAGE INDEX (vibration ; } d=ns^{6.4})$$

	Life			ess	
time/min (n)	2×10^6		time/min (n)	5	
g's rms (s)	1		g's rms (s)	6.00	

$$ess/life = 23.8\ \%$$

Thermal and Vibration Damage Index

The allowable percentage of useful life consumed by ESS is dependent on the particular application. The procedure described may require modifications based on individual design specifics and/or susceptibilities. Care should be taken when using the damage index. The index may be misleading for short term life items. From MIL-HDBK-344A, 16 August 1993.

SECTION 6

C.3 Lessons Learned

The following excerpts are questions that normally arise during ESS discussions and some experience learned during ESS implementation. The information for this section was taken from the following publications and is recommended reading.

APPLICABLE DOCUMENTS

DOD-HNBK-344/A	*Environmental Stress Screening (ESS) of Electronic Equipment*
MIL-HNBK-217	*Reliability Predictions of Electronic Equipment*
MIL-HNBK-785	*Reliability Program for Systems and Equipment Development and Production*
RADC-TR-81-86	*Environmental Stress Screening (ESS)*
RADC-TR-81-87	*Environmental Stress Screening (ESS)*
RADC-TR-81-130	*Environmental Stress Screening (ESS)*
RADC-TR-81-149	*Environmental Stress Screening (ESS)*
MIL-HNBK-338	*Electronic Reliability Design Handbook*
NAVMAT P-9492	*Navy Manufacturing Screening Program*
NAVSEA Notice 3900	*Electronic Hardware Stress Screening*
MIL-STD-2164 (EC)	*Environmental Stress Screening Process for Electronic Equipment*
Tri-Service ESS Guidelines	*5/92*

....

DETECTION EFFICIENCY

The Detection Efficiency (De) values can be found in Table 3-2 DOD-HNBk-344/A

Table 3-2 Approximate Values of Detection Efficiency for Various Test Types

Level of Assembly	Test Type	Detection Efficiency
	Production Line GO-NOGO Test	0.85
Assembly	Production Line In-Circuit Test	0.90
	High Performance Automatic Tester	0.95
	Performance Verification Test (PVT)	0.90
Unit	Factory Checkout	0.95
	Final Acceptance Test	0.98
	On-Line Performance Monitoring Test	0.90
System	Factory Checkout Test	0.95
	Customer Final Acceptance Test	0.99

DOD-HNBK-344

..

Types of Latent Defects

Application Of power, exercising and monitoring equipment performance continuously during the screen will greatly enhance test detection efficiency. Latent defects that are precipitated to failure by stress screening can be categorized into 3 types:

Type 1: HARD FAILURE; Physical defects that are readily transformed from an inherent weakness to a hard failure by the stress screen and remain in the equipment for post test detection. The usual yield of detection is 20 to 50 % of the latent defects.

Type 2; PHYSICAL DEFECT; Physical defects that manifest as failures only while under thermal or mechanical stress (i.g. intermittent caused by a cold solder joint or solder creep).

John J. Quinn

Type 3: FUNCTIONAL DEFECT; Defects that manifest as performance failures or anomalies only while under thermal or mechanical stress. (e.g. timing problems or voltage limits exceeded).

The type 1 defects are readily detected by post screen tests of sufficient thoroughness. Type 2 and type 3 defects require through and continuously monitored tests so that they can be detected. Type 3 defects, which include problems such as timing, part parameter drift with temperature or tolerance build-up can only be detected with powered and monitored tests. Type 2 and type 3 defects can comprise 50 to 80 % of the latent defects present in the equipment.

DOD-HNBK-344

....

ESS Algorithms for HP-20 Calculator

THERMAL CYCLE						RANDOM VIBRATION		
←	PRGM					LBL	B	
→	CLPRGM	Range in ° > STO 1				RCL	4	
		# of Cycles> STO 2						
→ ↓	LBL	Ramp in ° > STO 3				√	1.71	=
		A						
1	eˣ					×	RCL	5
+	RCL	3	=			×	0.0046	=
ln						+/-		
yˣ	3	=				eˣ		
×	RCL	2	=			STO	0	
STO	0					C		
C						1	.	
RCL	1					RCL	0	=
+	0.6	=				RTN		
×	RCL	0				To run program, enter g's rms into R4		
×	0.0017	=				and time in minutes into R5.		
+/-						Press:		
eˣ						XEQ	B	
STO	0							
C								
1	.							
RCL	0	=						
→	RTN							
To run program, enter values into r1,r2 & r3.								
Press	XEQ	A						

tceq.xls & rveq.xls J^2 algorithm

John J. Quinn

CONSTANT TEMPERATURE

←	PRGM	
→	LBL	C
RCL	6	
+	0.6	
y^x	0.6	=
×	RCL	7
×	0.0017	
+/-	e^x	
STO	0	
C		
1	-	
RCL	0	=
→	RTN	

To run program, enter temp range in R6 and
time in hrs in R7.
Range = deg C - 25 (ambient).
PRESS;

XEQ	C

cteq.xls   J2 algorithm

Example of ESS as a step Function rather than 1 screen.

Thermal Cycle:

120 = range

3 cycles

5° C/ramp

* If done three times

$1 - (1 - Tc1) (1 - Tc2) (1 - Tc3) = .9009$

* If done with nine cycles = .9009

Random Vibration:

6 g's rms

10 minutes

* For 10 minutes SS = .6265

* For 30 minutes , 3 ; 10 minute vibs = .9479

* For 30 minutes, 1 ; 30 minute vibe = .9479

J^2 Example

....

STRESS LEVELS

NOTE; FACTORY ESS STRESS LEVELS SHOULD ALWAYS BE
HIGHER THAN THE APPLICATION STRESS!
                                        DOD-HNBK-344

....

THERMAL CYCLING THEORY

It is the number of thermally induced stress reversals (minimization of soaks), the temperature extremes and the thermal rate of change of the hardware which are the principal parameters associated with disclosure of thermally sensitive manufacturing defects. The thermal of the thermal

John J. Quinn

chamber air is irrelevant. Chamber air temperature overshooting is an acceptable method of increasing the temperature rate of change.

TE 000-AB-GTP-020

....

Random Vibration Level for Screens

It is not necessary for the Random Vibration Stress Screen to exceed .04 g^2/Hz! The principal parameters in Random Vibration are the number of axis that are vibrated, the response of the equipment to the acceleration spectrum (AS) and the duration (time) of the vibration!

TE 000-AB-GTP-020

....

Sanders Assoc. Inc General Summary Notes

o Temperature and Vibration stresses are the two most effective screens!

o Hardware reliability improvements (MBTF) by factors of 3 to 5 (sometimes more) are routinely achieved!

o Screening should be performed at all levels of assembly starting at the lowest level possible!

o Environmental stress screening need not simulate the use environment, however, it does need to be adequate to expose these defects which could reasonably be expected to cause failures in the field!

o Temperature cycling tends to expose defects weakened by vibration screening but which did not actually fail under vibration stress. The overwhelming opinion is that temperature cycling should always be performed after vibration screening. If temperature cycling is performed, followed by vibration screening, then additional temperature cycling is necessary after the vibration screen !

o Many problems are conditional, that is, they appear only while stress is being applied, and vanish when the stress (temperature or vibration) is removed. It is important therefore that some degree of performance verification be performed during the screening process !

o A thermal survey and vibration survey must be performed on at least one sample to characterize the stress response of the equipment. The information obtained provides the basis for the detailed screen design! Stress responses are the key factors, not merely the screen inputs!

o The recommended screening sequence is: random vibration, temperature cycling, random vibration and thermal cycling!

DAMAGE INDEX

$$D = N S^b$$

Temperature cycling:
$b = 2.5$

	Life	ESS
N = # of cycles	$N1 = 7300$	$N2 = 50$
S = Temp range $^\circ$ C	$S1 = 30$	$S2 = 120$

% of useful life consumed by ESS = 21.9 %

Random vibration:
$b = 6.4$

	Life	ESS
N = Duration (minutes)	$N1 = 2E+o6$	$N2 = 5$
S = Level (grms)	$S1 = 1$	$S2 = 6$

% of useful life consumed by ESS = 27.3 %

DOD-HNBK-344

John J. Quinn

TEMP vs. VIB

It should be noted that, in practical terms, as many additional temperature cycles as are necessary may be applied without affecting the equipment's useful life.

Unlike thermal cycling, the maximum time that a unit can be exposed to the specified spectrum of random vibration, without significantly affecting its useful life, is severely limited.

MIL-STD-2164

....

TEST WHILE PERFORMING ESS

ESS that does not employ testing during stress application is relatively ineffective. It is also the reason why Random Vibration stress should be followed by Thermal Cycling.

DOD-HNBK-344

....

SHAKE THEN BAKE

The best screen for electronic hardware is to perform a printed wiring assembly (PWA) level thermal cycle stress screen before random vibration and a higher indenture level thermal screen after random vibration. The thermal screen prior to random vibration pre-stresses potential defects which can then be surfaced more efficiently by random vibration. Random vibration can also condition some defects just to the point of failure and a subsequent thermal screen with performance monitoring complies the defect identification.

TE 000-AB-GTP-020

....

VIBRATION TIME

The requirement for ten minutes in each axis is derived from studies that demonstrate maximum screening effectiveness is obtained in eight to ten minutes.

TE 000-AB-GTP-020

THERMAL & VIB COMBINED

When thermal cycling and random vibration are combined, random vibration shall be required towards the end of thermal cycling if combined at the printed wiring assembly level, or towards the beginning of thermal cycling if combined at a higher indenture level.

TE 000-AB-GTP-020

HARD FAILURES

If all of the failures that occurred were "hard" failures, performance monitoring might not be necessary. Unfortunately, many failures that occur in electronic hardware are "conditional" failures (intermittent). That is, they can only be detected while the thermal or vibration stress is being applied. Once the environmental stress is removed, the failure disappears.

TE 000-AB-GTP-020

POWER ON/OFF

A minor problem that occurs is interpreting the power on requirement as meaning power must be applied all the time during the thermal screen. Very poor screening decisions can result from this misunderstanding. For programs in full scale development, test equipment and test plans/procedures must be designed with power on/off cycling as a requirement at an indenture level above the PWA level. Power on/off cycling can be combined with thermal cycling for maximum results. It is strongly recommended that power be off during the cooling portion of the thermal cycle, and during any portion that is above or below the equipment operating specification limits.

TE 000-AB-GTP-020

<hr/>

DEFECTS REMAINING EXAMPLE

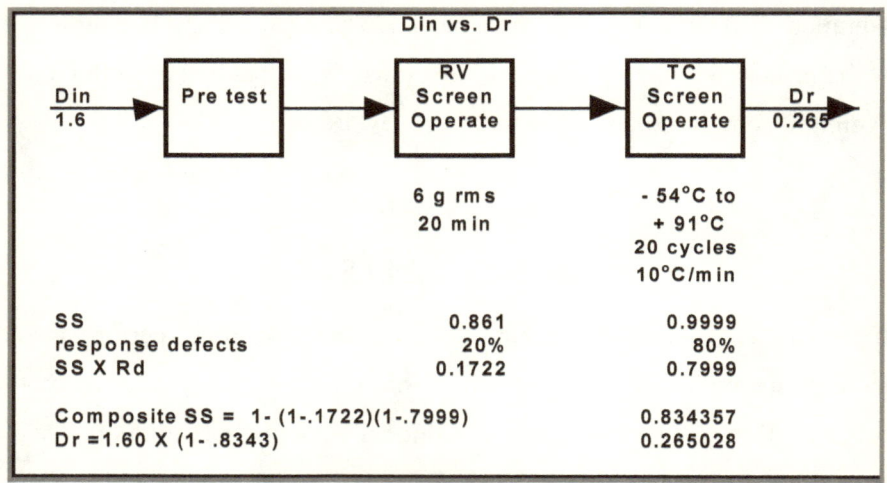

drdin.xls

Defects in is given as 1.6 and is pre tested at ambient to ensure all "patent" defects are detected. The first screen is random vibration using the NAVMAT P-9492 profile (notched). The vibration screen is one axis (perpendicular to the PWA) for 20 minutes. This yield 6 g's rms, and a Screening Strength of .861. A vibration screen usually detects 20 % of the total "latent" defects, this reduces the Screening Strength (SS) to .1722. The next screen is 20 cycles of thermal cycling, a temperature range of

145°C and a ramp of 10°C/minute. The SS is calculated as .9999, with a detection percentage of 80 %. The SS is now reduced to .7999. The composite SS of the two screens is .8343 and the defects remaining (passed on to the next level) is .2650. It should be pointed out that both screens are operating and therefore do not have to be reduced by 50 %.

J^2 Example

....

TEMP & VIB

Here's another example of comparing various stresses to get the same Screening Strength (SS).

o Ten Thermal cycles = .9127
o Ten continuous cycles or two (5) cycles ? Makes <u>NO</u> difference.
10 cycles yields a SS = .9127

2; 5 cycles yield-5cycles SS=.7046.Sscomp=$1-(1-.7046)^2$ = .9127

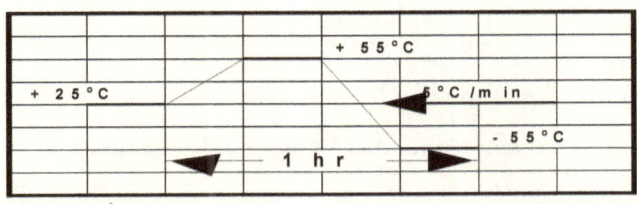

examp.xls

o Hi Temp Soak – 55 – 25 = 30oC Range for 75 hours = .6295
o Lo Temp Soak – 25°C to – 55°C = 80°C Range for 61 hours = .7640
o Hi & Lo Composite = 1-(1-.6295)(1- .7640) = .9126
o If just Hi Temp is used – 185 hours @ +55°C
o If just Lo Temp is used – 105 hours @ -55oC

CONCLUSION:

Thermal Cycling takes 10 hours, Hi and Lo Soaks take 136 hours. This is a 13:1 Ratio.

TEMP & VIB (con't)

VIBRATION COMPARISONS

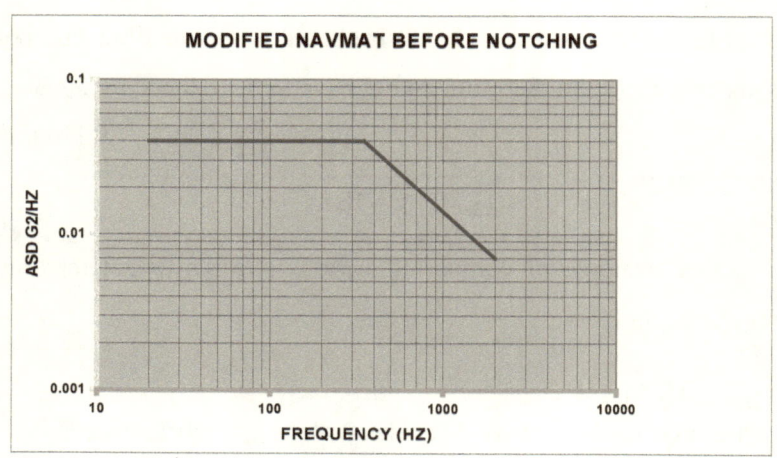

BREAK POINTS	
Freq	ASD
20	0.04
350	0.04
2000	0.007

nvmatm.xls

o SS for random vibration (6 g rms, for 15 minutes) = .7718

o SS for swept sine vibration (6 g's, for 440 minutes [7.3 hrs]) = .7717

o SS for single frequency sine vibration (6 g's, for 1307 minutes [21.8 hrs]) = .7719

....

BENEFITS OF ESS

o Reduced overall life cycle cost.

o On-time deliveries.

o Improved performance after delivery.

o Improved user confidence and/or satisfaction.

o Reduced support cost.

o Improved readiness.

....

DYNAMIC ESS

A viable ESS program must be <u>DYNAMIC</u> – the screening program must be actively managed, and tailored to the particular characteristics of the equipment being screened. This includes conducting a survey to determine the mechanical and thermal characteristics of the equipment and refining the screening profiles as more information becomes available and/or designs, processes and circumstances evolve.

Tri-Service ESS Guidelines, 5/92

....

ESS FAILURES

ESS is part of a viable engineering development, manufacturing, corrective action and overhaul process rather than a test in the normal accept/reject sense. Those participating in the effort, including the contractor, should never be led to believe that a " failure" is bad and would be held against them. <u>ESS is intended to stimulate defects, not to simulate the operating environment</u>, and therefore, factory "failures" are encouraged. The root cause of ESS failures need to be found and corrected before there is a complete process.

Tri-Service ESS Guidelines, 5/92

....

ESS PLANNING

It is imperative that ESS resources, training requirements and detailed plans (including levels of assembly and defined profiles) are in place when production begins. Therefore, it is desirable to reach this state during Engineering & Manufacturing development, so that hardware for Qualification and Reliability growth testing is of higher quality and can be screened (to prevent failures that are not design related). This implies that experimentation and planning should begin early. The cost of rework in manufacturing escalates by orders of magnitude as the assembly process proceeds from piece part level to printed circuit board/module, unit, system and to the user. Finding defects at the lowest possible level of assembly will tend to minimize rework costs by reducing corrective action time. However, some flaw types manifest themselves only at the higher level of assembly. Tailoring the screen to vibration and thermal characteristics of the hardware coupled with defect population at each level of assembly is essential.

Tri-Service ESS Guidelines, 5/92

THERMOTRON TEMPERATURE CONVERSION CHART

-350 to -70°F		-69 to -15°F		-12 to 44°F		46 to 101°F		102 to 168°F		159 to 610°F	
C°	F°	C°	F°	C°	F°	C°	F°	C°	F°	C°	F°
-212	-350	-56.1	-69	-24.4	-12	7.22	45	38.9	102	70.6	159
-207	-340	-55.6	-68	-23.9	-11	7.78	46	39.4	103	71.1	160
-201	-330	-55	-67	-23.3	-10	8.33	47	40	104	71.7	161
-196	-320	-54.4	-66	-22.8	-9	8.89	48	40.6	105	72.2	162
-190	-310	-53.9	-65	-22.2	-8	9.44	49	41.1	106	72.8	163
-184	-300	-53.3	-64	-21.7	-7	10	50	41.7	107	73.3	164
-178	-290	-52.8	-63	-21.1	-6	10.6	51	42.2	108	73.9	165
-173	-280	-52.2	-62	-20.6	-5	11.1	52	42.8	109	74.4	166
-167	-270	-51.7	-61	-20	-4	11.7	53	43.3	110	75	167
-162	-260	-51.1	-60	-19.4	-3	12.2	54	43.9	111	75.6	168
-156	-250	-50.6	-59	-18.9	-2	12.8	55	44.4	112	76.1	169
-151	-240	-50	-58	-18.3	-1	13.3	56	45	113	76.7	170
-145	-230	-49.4	-57	-17.8	0	13.9	57	45.6	114	77.2	171
-140	-220	-48.9	-56	-17.2	1	14.4	58	46.1	115	77.8	172
-134	-210	-48.3	-55	-16.7	2	15	59	46.7	116	78.3	173
-128	-200	-47.8	-54	-16.1	3	15.6	60	47.2	117	78.9	174
-123	-190	-47.2	-53	-15.6	4	16.1	61	47.8	118	79.4	175
-118	-180	-46.7	-52	-15	5	16.7	62	48.3	119	80	176
-112	-170	-46.1	-51	-14.4	6	17.2	63	48.9	120	80.6	177
-107	-160	-45.6	-50	-13.9	7	17.8	64	49.4	121	81.1	178
-101	-150	-45	-49	-13.3	8	18.3	65	50	122	81.7	179
-95.6	-140	-44.4	-48	-12.8	9	18.9	66	50.6	123	82.2	180
-90	-130	-43.9	-47	-12.2	10	19.4	67	51.1	124	82.8	181
-84.4	-120	-43.3	-46	-11.7	11	20	68	51.7	125	83.3	182
-78.9	-110	-42.8	-45	-11.1	12	20.6	69	52.2	126	83.9	183
-73.3	-100	-42.2	-44	-10.6	13	21.1	70	52.8	127	84.4	184
-72.8	-99	-41.7	-43	-10	14	21.7	71	53.3	128	85	185
-72.2	-98	-41.1	-42	-9.44	15	22.2	72	53.9	129	85.6	186
-71.7	-97	-40.6	-41	-8.89	16	22.8	73	54.4	130	86.1	187
-71.1	-96	-40	-40	-8.33	17	23.3	74	55	131	86.7	188
-70.6	-95	-39.4	-39	-7.78	18	23.9	75	55.6	132	87.2	189
-70	-94	-38.9	-38	-7.22	19	24.4	76	56.1	133	87.8	190
-69.4	-93	-38.3	-37	-6.67	20	25	77	56.7	134	88.3	191
-68.9	-92	-37.8	-36	-6.11	21	25.6	78	57.2	135	88.9	192
-68.3	-91	-37.2	-35	-5.56	22	26.1	79	57.8	136	89.4	193
-67.8	-90	-36.7	-34	-5	23	26.7	80	58.3	137	90	194
-67.2	-89	-36.1	-33	-4.44	24	27.2	81	58.9	138	90.6	195
-66.7	-88	-35.6	-32	-3.89	25	27.8	82	59.4	139	91.1	196
-66.1	-87	-35	-31	-3.33	26	28.3	83	60	140	91.7	197
-65.6	-86	-34.4	-30	-2.78	27	28.9	84	60.6	141	92.2	198
-65	-85	-33.9	-29	-2.22	28	29.4	85	61.1	142	92.8	199
-64.4	-84	-33.3	-28	-1.67	29	30	86	61.7	143	93.3	200
-63.9	-83	-32.8	-27	-1.11	30	30.6	87	62.2	144	98.9	210
-63.3	-82	-32.2	-26	-0.56	31	31.1	88	62.8	145	100	212
-62.8	-81	-31.7	-25	0	32	31.7	89	63.3	146	104	220
-62.2	-80	-31.1	-24	0.56	33	32.2	90	63.9	147	110	230
-61.7	-79	-30.6	-23	1.11	34	32.8	91	64.4	148	116	240
-61.1	-78	-30	-22	1.67	35	33.3	92	65	149	121	250
-60.6	-77	-29.4	-21	2.22	36	33.9	93	65.6	150	127	260
-60	-76	-28.9	-20	2.78	37	34.4	94	66.1	151	132	270
-59.4	-75	-28.3	-19	3.33	38	35	95	66.7	152	138	280
-58.9	-74	-27.8	-18	3.89	39	35.6	96	67.2	153	166	330
-58.3	-73	-27.2	-17	4.44	40	36.1	97	67.8	154	216	420
-57.8	-72	-26.7	-16	5	41	36.7	98	68.3	155	238	460
-57.2	-71	-26.1	-15	5.56	42	37.2	99	68.9	156	243	470
-56.7	-70	-25.6	-14	6.11	43	37.8	100	69.4	157	249	480
		-25	-13	6.67	44	38.3	101	70	158	321	610

tempcon.xls

ESS at a particular level cannot uncover flaws that are not introduced until the next level. Generally, this dilemma is usually controlled by performing ESS at each major functioning level in the manufacturing process consistent with an assessment of defect population at each level of assembly.

Vibration is the area of stressing that normally precipitates latent assembly flaws caused by the undesired relative motion of parts, wires, structural elements, etc. as well as mechanical flaws that lead to propagating cracks. Effective screening requires large, rapid temperature changes and broadband vibration. Such thermal cycling is used for the detection of assembly flaws that involve installation errors or inadequate chemical or mechanical isolation or bonding. Under rapid thermal cycling, different thermal expansion takes place without sufficient time for stress relief, and this is a major mechanism for precipitating latent defects to detectable failures. If random vibration and thermal cycling are conducted sequentially, random vibration should be done first. The technical risks and costs are summarized below:

Tri service ESS guidelines

	TEMPERATURE CYCLING								
	ESS CONDITIONS/TRADEOFFS						RISKS/EFFECTS		
Level of Assembly	Power Applied		IO [1]		Monitored [2]		ESS Cost	Technical		Comments
	yes	no	yes	no	yes	no		Risk	Results	
		X		X		X	low	low	poor	conduct pre/post ESS functional test & often
PCB	X		X		X		high	lower	better	prior to conformal coating.
	X			X	X		highest	lowest	best	
	X		X		X		highest	lowest	best	If circumstances permit ESS at only one level of
UNIT	X			X	X		lower	higher	good	assembly, implement at unit level.
		X		X		X	lowest	highest	poor	
SYSTEM	X		X		X		highest	see comment		Most effective ESS at system level is short duration random w/o too long to intercept defects resulting from system integration.
	Ground rule: screen at lowest level of assembly consistent with defects present at that level of assembly !!									
	RANDOM VIBRATION									
	ESS CONDITIONS/TRADEOFFS						RISKS/EFFECTS			
Level of Assembly	Power Applied		IO [1]		Monitored [2]		ESS Cost	Technical		Comments
	yes	no	yes	no	yes	no		Risk	Results	
		X		X	X		highest	low	good	Random w/o can be effective @ pcb if surface
PCB	X			X	X		high	high	fair	mount technology is used, pcb has large com pon,
		X		X		X	low	highest	poor	pcb is multilayer & can't be screened at higher level
	X		X		X		highest	low	optimum	RV effective at level of assy w/min interconn flaws & most
UNIT	X			X	X		low	higher	good	susceptable to power on w/ITO ESS. Pow er on
		X		X		X	lowest	highest	poor	without ITO reasonable effective.
SYSTEM	X		X		X		low	low	good	Cost is low power on & ITO testing due to acceptance test. ESS at this level needed if there is expected flaws.

Notes:
1 = I/O - Equipment fully functional with normal inputs and outputs.
2 = Monitored - during screen monitoring key points to assure proper equipment operation.

essrisk.xls

SURVEYS

To avoid potential fatigue or peak level damage due to resonance, some level reduction of the input spectrum may be done at points of severe resonant frequencies which result in amplification of the applied stress level by a factor of 3 db or more. Notching (but not notching out) may be permitted, but should not be the exception.

Temperature cycling screens also have to be tailored to each specified equipment and are equipment unique. Differences in components, materials and heat dissipation lead to variations in the thermal stresses throughout the item.

Tri service ESS guidelines

WORKMANSHIP vs. PARTS

Since the primary purpose of ESS is to precipitate latent problems associated with the manufacturing processes, its effective use is predicated on good design with quality parts. Historically, ESS results show that failures due to workmanship are approximately two thirds of the total with the other third due to bad parts and poor design.

Tri service ESS guidelines

TEST, ANALYZE and FIX (TAAF) PROGRAMS

TAAF reliability growth testing programs are used extensively by the services to identify and correct design deficiencies on new systems while still in the engineering & manufacturing development phase. ESS should precede the formal TAAF testing. This helps minimize the occurrences of failures not related to design inadequacies. Unrelated failures tend to

John J. Quinn

retard the TAAF processes, lengthen the test time and increase the total costs of the TAAF tests.

Tri service ESS guidelines

....

FRACAS

One of the best practices in successful system development efforts is the proper implementation of a failure reporting and corrective action system (FRACAS). As defined in the military standards, FRACAS is a "Closed-loop system for initiating reports, analyzing failures and feeding back corrective actions into the design, manufacturing and test processes." Thus, ESS is an essential tie to the design and manufacturing processes during development and to statistical process control (SPC) of the manufacturing processes during production and depot repair.

Tri service ESS guidelines

....

SAMPLING vs. 100% SCREENING

If it can be demonstrated that the decline in ESS failures is indeed due to improvements, and not to manufacturing changes that make the ESS conditions ineffective, suspension of 100% ESS may be considered.

However, monitoring should be instituted to make sure that the improvements remain effective. The best way to accomplish this is to continue ESS on a sample basis, with reversion to 100% ESS on evidence of loss of process control. One hundred percent ESS also should resume when processes, parts or sources are changed and after production breaks. In most military contracts the production quantities are not sufficient to justify the effort necessary to go from 100% screening to a sampling procedure.

Tri service ESS guidelines

102

CHAMBER AIR FLOW

In thermal stress screening, the rate of change of temperature is as important as the temperature extremes. The faster the rate of change, the more effective the temperature stress screen. But it is the individual components that must experience a particular rate of change of temperature extremes. To attain the appropriate temperature rate of change and temperature extremes of the item being screened, there are several things that the screen designer may be able to do:

o Allow the ESS chamber to "overshoot" the temperature parameters. Overshooting is a method of achieving an increases temperature rate of change and higher/lower temperature extremes when the chamber air temperature exceeds the upper and lower screening temperature limits for a controlled period of time. Controlled overshooting is permissible and encouraged as an excellent method of achieving higher temperature rtes of change, thereby increasing screen effectiveness. To avoid overstress, the temperature of the part with the smallest thermal mass should be monitored with a thermocouple.

o Remove the protective covers of the equipment thus allowing the chamber air flow to more easily reach the individual components.

o As the thermal mass increases, the air flow becomes more restricted. To compensate for this, an air circulating system (fans) can be installed to direct air to the areas of the unit with the highest thermal mass, thus causing the components to experience a much greater temperature rate of change.

Tri service ESS guidelines

REPEATED SCREENING

Repeated application of screens after correction of ESS flaws can very easily begin to use up significant useful life and to initiate rather than precipitate flaws. To avoid such counter-productive screening, the following guidelines are recommended:

o After repair of failure during first operating vibration screen, complete remaining duration of screen, or five minutes, whichever is greater.

o After repair of failure during first non-operating vibration screen, repeat screen at full level and 50% duration.

o After subsequent repairs and/or modifications, repeat screen at − 3 db level (70% g rms) for 50% duration.

o Total number of vibration exposures not to exceed five.

o If failure is detected and repaired during the initial thermal cycling screen, then the balance of the cycles scheduled, or a minimum of three should be run.

o After subsequent repairs and/or modifications, run one complete thermal cycling screen.

The guidelines above should be used in conjunction with assessment of the appropriate amount of re screening which takes into account the nature of the repair/modification, the amount of tear down, rework and re assembly involved and the chance for introducing workmanship flaws. Such assessments are appropriately made through Corrective Action Boards/Failure Review Board actions.

Tri service ESS guidelines

....

VIBRATION SCREEN DEVELOPEMENT CONSIDERATIONS

METHOD	PRO	CON
A. Vibration Survey	Two techniques to determine spectral responses for tailoring.	General survey technique requires spectral analysis equipment.
B. Step-Stress Tests	Stright forward emperical method if performed by experienced engineers. May provide equipment with increased durability. Defines item design limits. Ideal for existing and developing technology.	Some risks of overstress if design limitation is unknown.
C. Fault Replication Tests	Good supplement to method B.	Lack of hardware with repeatable failure modes. Difficulty in "seeding" hardware realistically.
D. Heritage Screen	Minimum developement resources required. Most easily justified to approving authority.	Transparent dissimilarities may yield inadequate or damaging screen.

devcon.xls

C.☐   Vibration Survey (Simplified and General)

Simplified – If the response values are within +/- 3 db then no tailoring is necessary. Where the response values differ from the excitation level by more than +/- 3 db, then tailoring is warranted. A low level vibration[1] should be run to obtain responses by frequency at each location where the differences were more than +/- 3 db. With these data, the input acceleration spectral density can be tailored to obtain the desired output response of 6 to 10 grms throughout the unit being screened.

General – The development of a random vibration stress screen is predicted on tailoring the input to achieve an acceptable response throughout the unit being screened. A vibration survey is the most logical and straightforward way to determine these responses. The spectral responses from selected accelerometer sites (approximately 20 locations should suffice for mapping most items) identify the frequencies where

105

responses or damping occur. The input vibration level at appropriate frequencies can then be tailored to eliminate undesired high or low responses.

B. Step-Stress Test

Step-stress testing is an empirical procedure that can be used when resources for elaborate surveys, recording, analysis and technical support are limited. Due to the associated inaccuracies and risks, however, it's use must be approved by the government. The step-stress approach determines the "tolerance limit" or design capability of the hardware for the screen. By knowing this limit, a safe screening level can be determined and changed as required to obtain satisfactory screening results. The overall input level is tailored to the product. As in method A, the vibration survey test configuration should replicate the configuration for the proposed screen. The test item must be representative of the hardware to be screened. It should be permissible to accumulate vibration time on the test hardware. The fixture, slip-plate and head expander used for the survey should be the same as for the screen.

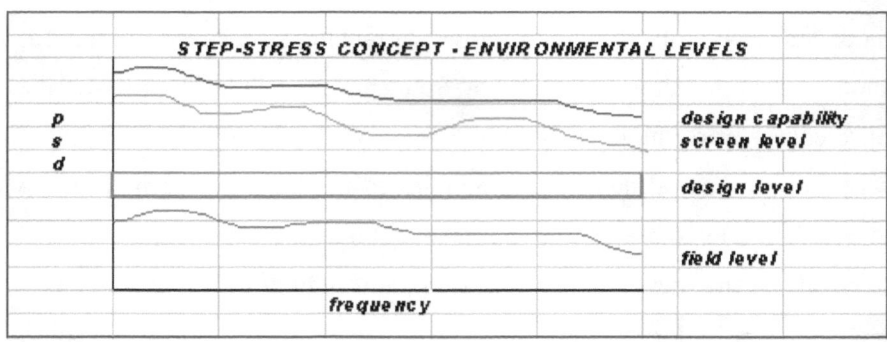

C. Fault Replication Tests

Fault replication testing increases the input screening level until known faults in the unit being screened are precipitated. These known faults may be recognized manufacturing problems or faults that have been

106

deliberately seeded. As in the case of method B, however, use of this method must be approved by the government. Method C can be used independently to establish a screening level or can be used in conjunction with method B. In either case, the steps involved are similar to those listed for method B except that the testing stops at the input level necessary to precipitate the known faults. A minimum of 10 faults should be available for replication to establish an effective screening level. Ideally, the faults should be "hidden" or latent, meaning that they would not be detected during functional tests. Representative faults to be used for seeding are:

o Loose hardware o Flawed components o cold solder joints
o Nicked component legs o Incorrect bonding o Removed bonding
o Intermittent switch o Insufficient solder o Connector partially seated
o Nicked wires o Fractured hardware

D. Heritage Screen
A heritage screen is a screen derived from recent successful experience on equipment of comparable design and manufacturing. Ideally, this experience would have been based on Method A. A heritage screen should only be considered if there are data substantiating its effectiveness. Government approval is required for method D.

Tri service ESS guidelines
1 = jj quinn – navmat profile @ .001 psd

TEMPERATURE CYCLE CHARACTERISTICS

A thermal survey evaluates the thermal response of various elements in the hardware to changes in the temperature of the chamber air. The results of the thermal survey will be experimental plots of the thermal responses, measured at critical elements of the hardware, to changes in the chamber air. The necessary temperature range and rate of change of the chamber air can then be identified for a desired response.

Tri service ESS guidelines

....

TEMPERATURE EXTREMES

The temperature extremes in a thermal cycle affect the effectiveness of the screen. The temperature range dictates the thermal stress/strain to which the hardware is subjected in each cycle. The number of cycles to failure varies inversely with the temperature range: the wider the range, the earlier the failure. By optimizing the temperature extremes, the screening profile designer can minimize the number of cycles required to precipitate flaws. Thus, the temperature extremes also affect the cost of the screen. The key to selecting the temperature extremes is to stress the hardware adequately to precipitate flaws without damaging good hardware. In practice, temperature ranges from a minimum of 90°C to a maximum of 180°C have been used. Maximum values are: 125°C for modules (usually − 50°C to 75°C), 110°C for units (usually − 40°C to 70°C) and 100°C for systems (usually − 40°C to 60°C). The following key factors should be considered for the extreme values:

o Storage temperature (high and low) limits of hardware such as the materials in printed circuit boards.

o Maximum operating temperatures of electronic parts.

o System requirements (jj quinn input).

Tri service ESS guidelines

....

RATE OF CHANGE OF TEMPERATURE

Consistent with this phenomenon, industry has found that increasing the temperature rate of change increases the screening strength up to a point.

However, the situation is more complicated for solder, which creeps at temperatures encountered in thermal stress screening. Creep, which has been identified as the major cause of solder joint failure, requires time to occur. If the temperature rate of change is too high, the thermal stress screening profile may actually be excessively benign for the purpose of precipitating defective solder joints to failure. If properly conducted, environmental stress screening to precipitate defective solder joints in a specific set of equipment should have to be performed at only one level of assembly.

The choice of temperature rate of change depends on the nature of the hardware and the flaws expected. A high temperature rate of change is expected to be the most effective for precipitating flaws in such elements as plated-through holes, where as a slow rate of change with long dwells at high temperature is expected to be the most effective for precipitating flaws in solder joints. In practice the temperature rate of change varies from 5°C/min to 20°C/min with the nominal values for module screening being 15°C/min to 20°C/min and for systems, 5°C/min to 10°C/min. The screening strength does not increase indefinitely with increasing temperature rate of change.

Tri service ESS guidelines

John J. Quinn

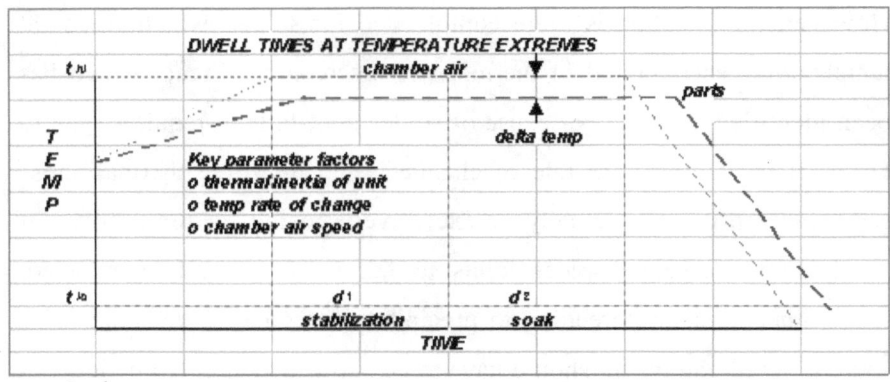

ttcyc2.xls

Stabilization

The stabilization time (d1) required for internal components to reach the ultimate chamber temperature (chamber temperature set point) has to be determined by the thermal survey. The choice of stabilization criterion affects the duration and thus the cost of the screen. The recommended stabilization criterion is: stabilization has occurred when the temperature of the slowest-responding performance-related elements in the hardware being screened are within 15% of the ultimate temperatures. The most important factors are the thermal inertia of the assembly being screened and the chamber air speed.

Soak

The soak period (d_2) serves two purposes. This period allows solder creep. The time required for solder to relax is on the order of five minutes. Furthermore, for screens in which the equipment is powered and monitored, the soak periods at the temperature extremes enable functional testing to be performed to detect failures which do not manifest themselves at ambient temperature. The recommended values of soak time (d_2) are as follows:

    o Unmonitored equipment – 5 minutes.

    o Monitored equipment – long enough for functional testing or 5 minutes, whichever is longer.

....

EQUIPMENT CONDITION

Detection of failures induced by the environmental stresses generally requires that the equipment be powered and monitored. Testing the equipment to detect failures should be done during application of environmental stress screening, otherwise intermittent failures will go undetected. Testing only before and after stressing results in high risk of letting the intermittent flaws remain. Some factors involved in deciding whether or not to have the equipment operating are as follows:

o A powered screen is more effective in precipitating flaws that a non powered screen. Powering produces temperature gradients in the hardware not present in a non powered equipment. The thermal stresses/strains resulting from these thermal gradients may precipitate flaws that escape in non powered screens.

o A powered and monitored screen may detect failures that escape in a non powered screen (intermittent failures). Failures that do not manifest themselves in testing at ambient conditions may show up in testing at high or low temperature or during vibration. An example is a broken connection in which the pieces are touching just enough to provide continuity in the absence of thermal/vibration stresses.

o A powered and monitored screen is more expensive than a non powered screen.

A power off screen at the printed circuit board level of assembly is recommended and effective for latent part defects*. However, it should only be considered if the pcb will see a powered screen at the next higher level of assembly.

John J. Quinn

ASSEMBLY LEVEL	EQUIPMENT CONDITION
Board	Non powered
Unit	Powered-Monitored
System	Powered-Monitored

X.□= hard failures (20 to 50 % of latent failures detected)

Tri service ESS guidelines

....

NUMBER OF CYCLES

As do the cycle characteristics, the choice of the number of cycles impacts the effectiveness and the duration and thus the cost of the screen . Thermal cycling produces stresses which induce alternate expansion and contraction. The stresses and strains are the highest at flaws because each flaw creates a stress riser that allows the stress to precipitate a flaw (i.e., latent defect) to hard (i.e. detectable-patent defect) failure. The cyclic loading causes the flaws to grow. Eventually they become so large that they cause a complete structural failure and thus an electrical failure. For example, a crack plated through hole eventually cracks completely around and causes an open circuit.

For solder, the physics of failures induced by thermal cycling is more complex than for materials such as aluminum and copper. The reason is that, at the temperatures encountered in electronics equipment, solder creeps. Creep has been identified as a major cause of solder joint failures. Solder creeps at a rate that increases with increasing temperature. Consequently, the number of cycles to failure of solder joints depends on other parameters as well as temperature range. The most severe thermal cycles for solder are those in which creep has sufficient time to occur. However, a screen should avoid unnecessarily inducing creep in solder joints.

o Be sure that the thermal survey and analyses have been completed to identify the most appropriate values of temperature range, temperature rate of change, dwell times and whether powered and monitored.

Tri service ESS guidelines

....

FAILURE FREE ACCEPTANCE TEST EXAMPLE

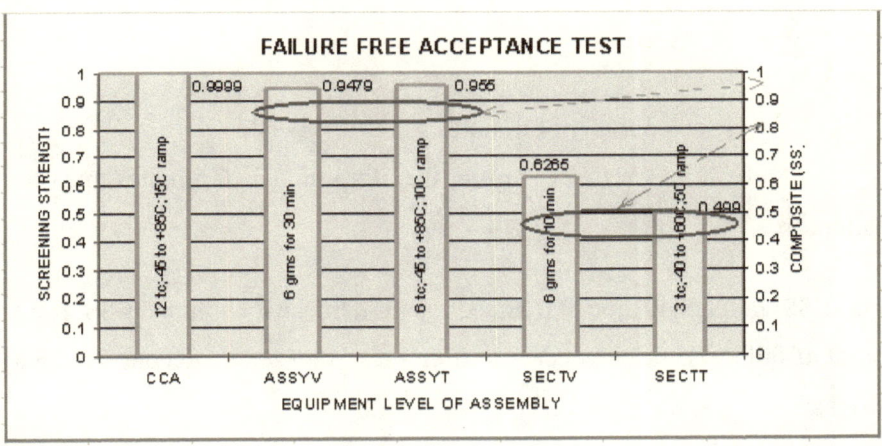

ffat1.xls

Using the above model, an example of FFAT will be demonstrated.

 o given – the above model and MBTF requirement of 500 hours.

Assembly Level:

 o Tc_{assy} = number of cycles X 1/1 + (ss) + K

 = 6 X 1 / 1 + (.95) + 0 = 3, therefore the <u>last 3 cycles must be</u> <u>failure free</u>.

 o Rv_{assy} = number of minutes X 1/1 + (ss) + K

= 10 X 1 / 1 + (.947) + 0 = 5, therefore the last 5 minutes must be failure free.

Section Level:

o Tc_{sect} = number of cycles X 1/1 + (ss) + K

= 3 X 1 / 1 + (.499) + 0 = 2, therefore the last 2 cycles must be failure free.

o Rvsect = number of minutes X 1/1 + (ss) + K

= 10 X 1 /1 + (.62) + 0 = 6, therefore the last 6 minutes must be failure free.

For ESS failures in non-ffat time, repair item and continue ESS from point of failure. If failure occurs in ffat time, repair item and continue ESS from start of ffat time.

	K FACTOR TABLE		
	MBTF	**K**	
	2000	2	
	1000	1	
	500	0	
	250	-1	
	100	-2	

kfact.xls

j2 quinn

....

SOLDER CREEP

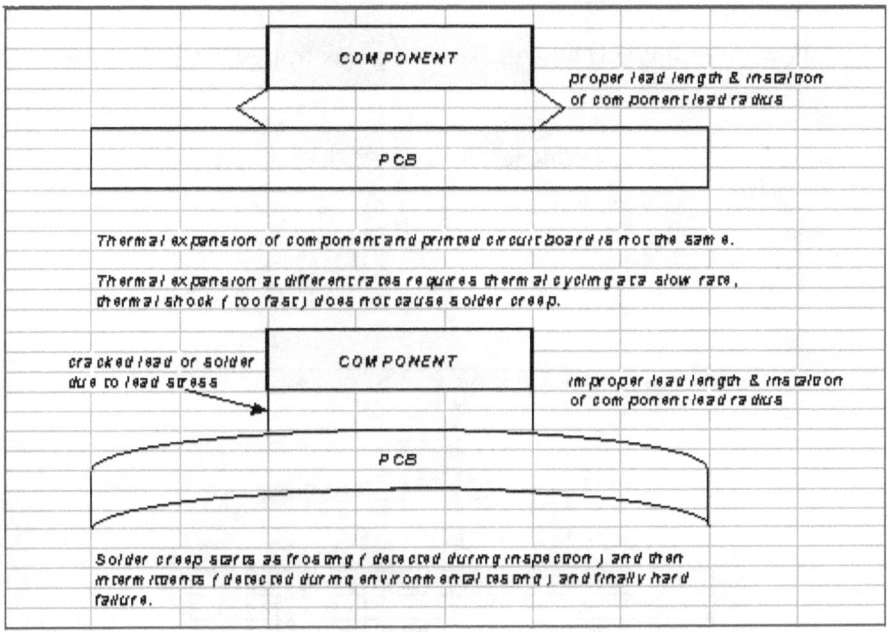

COMPONENT

proper lead length & installation
of component lead radius

PCB

Thermal expansion of component and printed circuit board is not the same.

Thermal expansion at different rates requires thermal cycling at a slow rate, thermal shock (too fast) does not cause solder creep.

cracked lead or solder
due to lead stress

COMPONENT

improper lead length & installation
of component lead radius

PCB

Solder creep starts as frosting (detected during inspection) and then intermittents (detected during environmental testing) and finally hard failure.

creep.xls

j2 quinn

CALCULATION OF GRMS

The following example is how to estimate g rms from vibration profiles:

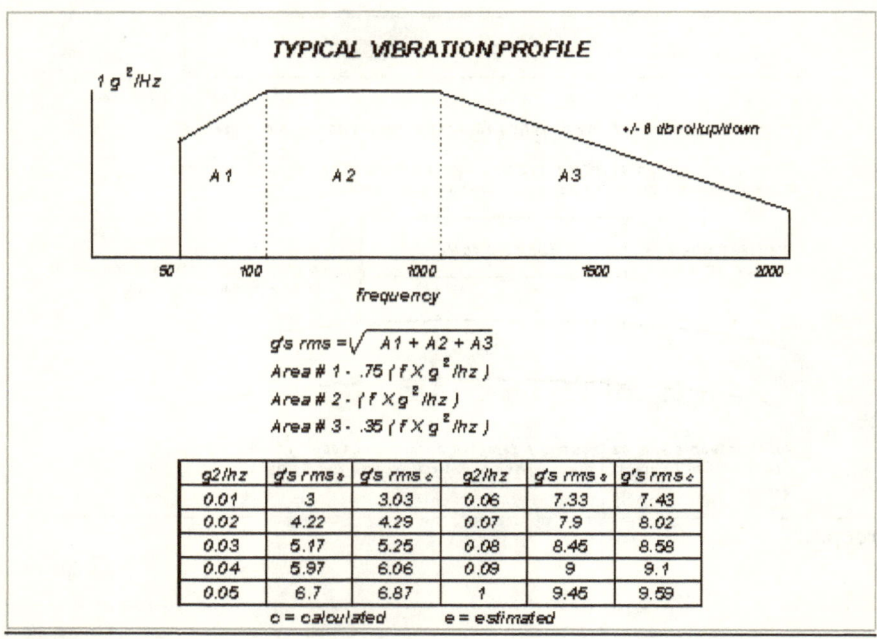

typpro.xls

The calculations for areas are:

o $A1 = pf1 / z1 \ [1 - (f1 / f2)^{z1}]$, where $z1 = r / 3 + 1$

$z1 = 6/3 + 1 = 3$

$A1 = 1.0 \ (100)/ \ 3 \ [1 - (50/100)^3] = 29.2 \ g^2$

o $A2 = f (g^2 / hz)$

$A2 = 1.0 \ (900) = 900 \ g^2$

o $A3 = pf1 / z2 \ [\ 1 - 1 / (f2 / f1 \ )^{z2}]$, where $z2 = r / 3 \ - 1$

116

$z2 = 6/3 - 1 = 1$

$$A3 = 1.0\ (1000)/1\ \{1 - [1\ /\ (2000/1000)^1]\} = 500\ g^2$$

$$grms = (29 + 900 + 50\ )^{1/2} = 37.8$$

Note: If $r = 3$, then $A3 = 2.3\ pf1\ \log f2\ /\ f1$

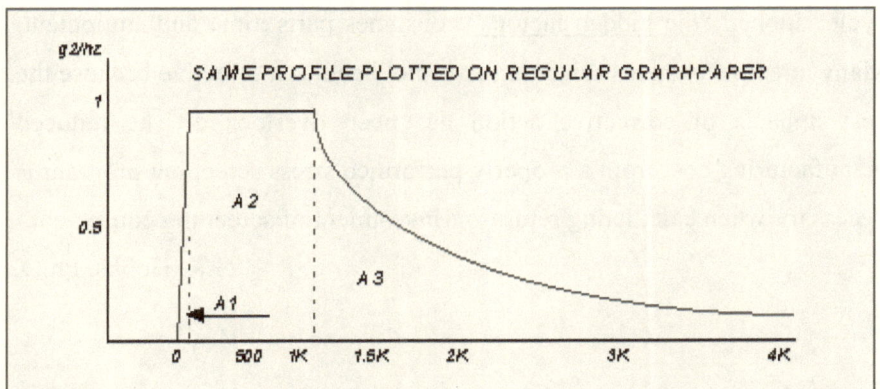

typpro2.xls

Tustin Institute of Technology

TESTING DEFINITIONS

Qualification testing, acceptance testing, MTBF testing and the screening process have very different goals, and therefore, procedures:

Qualification testing is performed to prove that a design can withstand at least one lifetime of service in the field.

Acceptance testing is a proof test to show that a particular specimen performs according to a specified test procedure. Defect stimulation is not intended in acceptance testing.

MTBF testing is intended to determine the failure rate in a defect free population. A flawed or non-screened population will provide a very

John J. Quinn

biased and erroneous estimate of the MTBF. Once the built-in flaws are removed, the MTBF will increase.

Stress screening is a process where defects are stimulated into being observable and therefore removable before acceptance testing and shipment. One objective of the screening process is to stop making product with the same or similar defects – to reduce work. The rework cycle, dubbed "the hidden factory" consumes parts ,time and equipment. Many attempts at stress screening have not been cost effective because the cost impacts of corrective action has been overlooked. The reduced manufacturing cost from a properly performed stress screening program is necessary when calculating return-on-investment of screening equipment.

G.K. Hobbs, Ph.D.

....

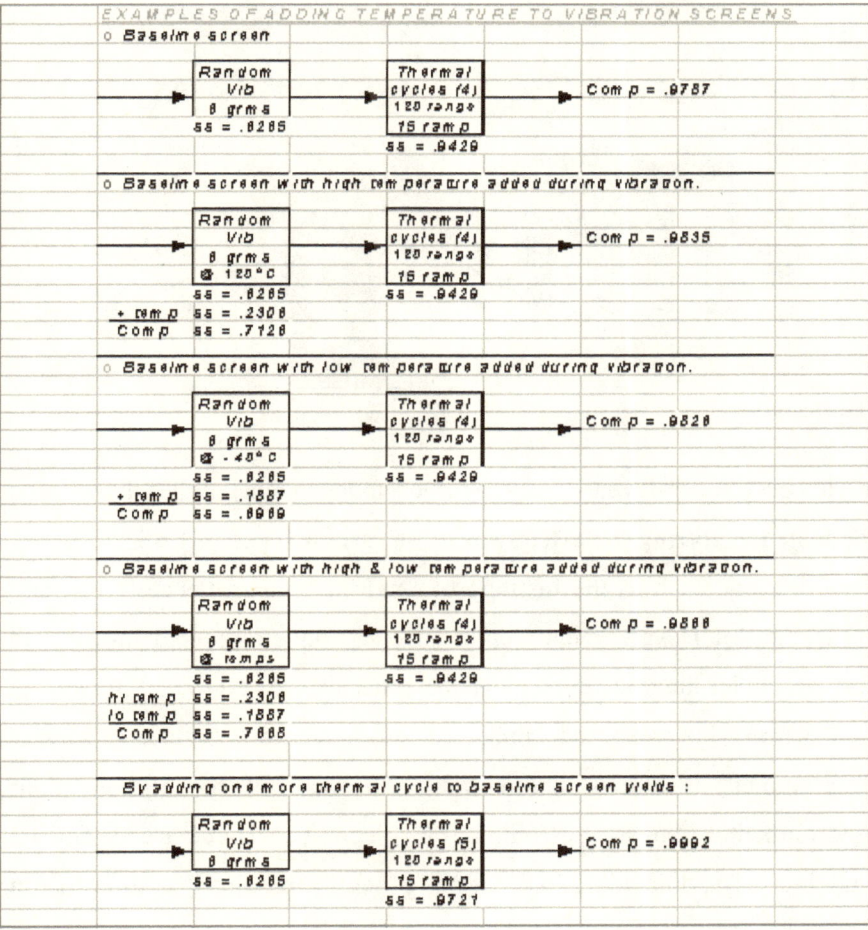

vibtemp.xls

QUESTION – Is it more cost effective to add temperature to the vibration screen rather than add one more temperature cycle to the temperature screen? I THINK NOT!

J2 quinn

....

John J. Quinn

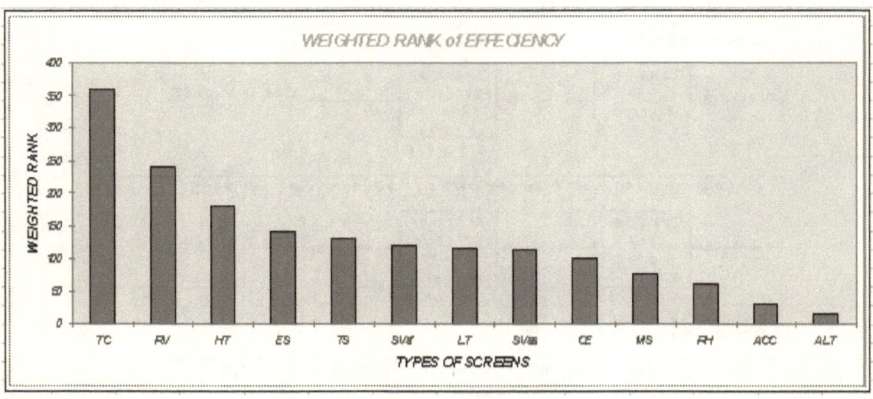

essrank.xls

In the past several years, a number of independent studies have been conducted analyzing the effectiveness of various types of ESS processes. Organizations such as the Institute of Environmental Sciences (IES) and the Rome Air Development Center (RADC) have compiled in-use data that concludes thermal cycling is the most effective type of screen. A comparison in weighted rank of effectiveness between various environments is shown above was published by the IES in their Environmental Stress Screening of Electronic Hardware (ESSEH) guidelines. Random vibration screens generally require less time to run than other ESS programs and are considered particularly effective in exposing mechanical defects, such as loose solder, improper bonds and PCB shorts. Thermal cycling when compared with random vibration (ranked second most effective), regularly detected an average of two-thirds more latent product defects.

Thermotron ESS handbook

120

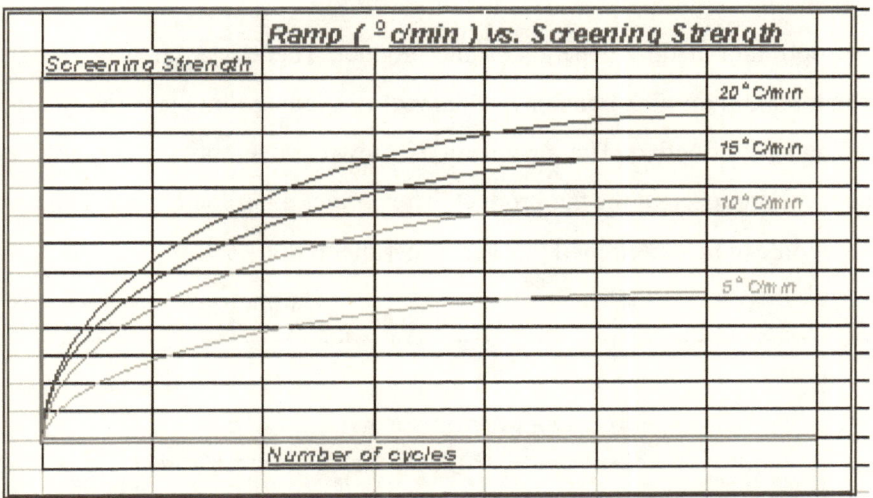

essair.xls

The chart above shows a comparison of various temperature change rates and their impact on screen effectiveness.

Thermotron ESS handbook

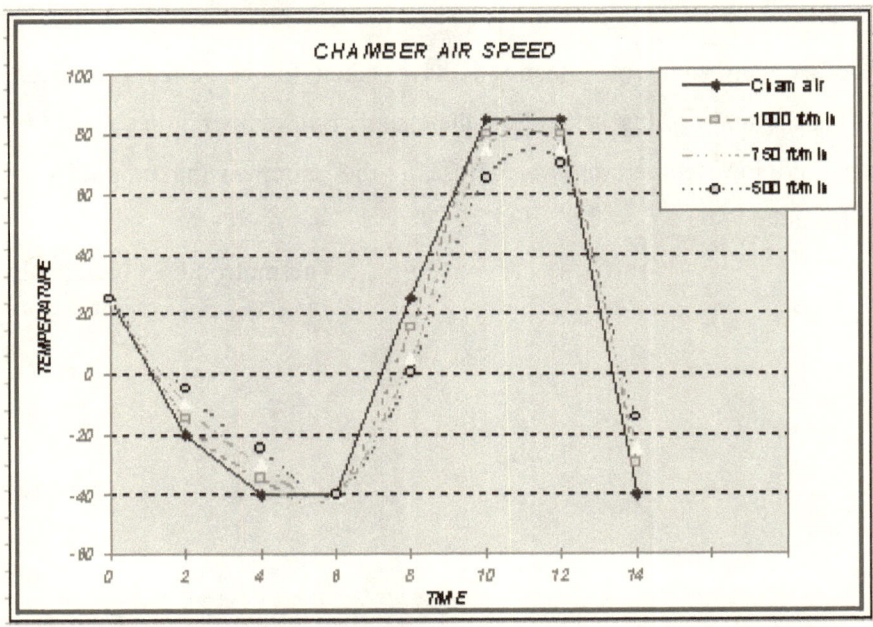

airspeed.xls

The optimum airflow depends on the product. There is an air velocity at which maximum heat transfer is obtained. Exceeding that air velocity can be counterproductive. The air velocities shown above may or may not be the most appropriate for the product. The correct air velocity rates and air direction can be determined through experimentation.

Thermotron ESS handbook

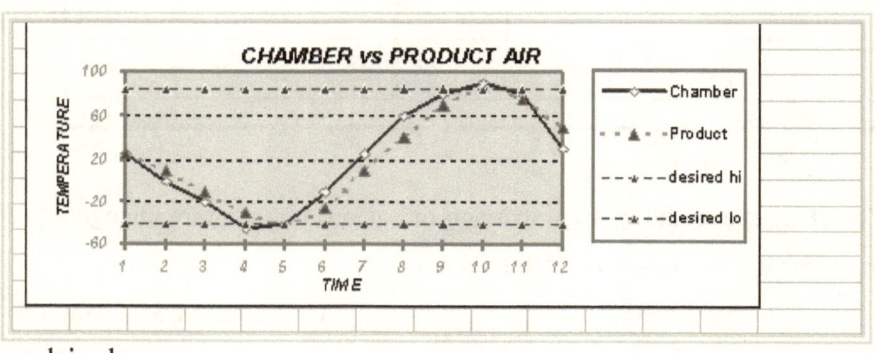

prodair.xls

Another step some manufacturers have taken to maximize the stress during thermal cycling is to adjust the chamber air temperature so that the high and low temperatures are close to the extremes the product can withstand.

Thermotron ESS handbook

Appendix A
Work sheet Calculations

John J. Quinn

A.1 Appendix A – Equations Tab

This excel worksheet will calculate the screening strength for "Thermal Cycling", "Constant temperature", "Random Vibration", "Swept Sine Vibration" and "Single Frequency Vibration".

There is also a "Temperature Conversion" work sheet.

For Temperature, the variables are "Range: - minimum temperature to maximum in degrees centigrade. "Cycles"- the number of temperature cycles that will be applied. "Ramp" is the change in temperature rate expressed in degrees centigrade per minute. The last veriable whether the Unit under Test will be operating during the screen. Remember the screen is only 20 to 50% effective if the unit is not operating. Non operating screens will only detect Type I" defects that are only detected during post ess electrical testing.

For Vibration, the variables are "g's rms" and "Time" in minutes. The same logic applies for operating vs. non operating during vibration screens. A vibration fixture is usually required for vibration to eliminate resonance's induced from the vibration shaker.

Both Temperature and Vibration screening requires a survey to be performed. That means that thermal couples and accelerometers must be installed to allow the engineer to record the thermal chamber characteristics and the vibration frequencies that may cause resonance. Notching out these resonant frequencies is a common tool.

A printout of the "ESS Work.XLS" file is on the following page.

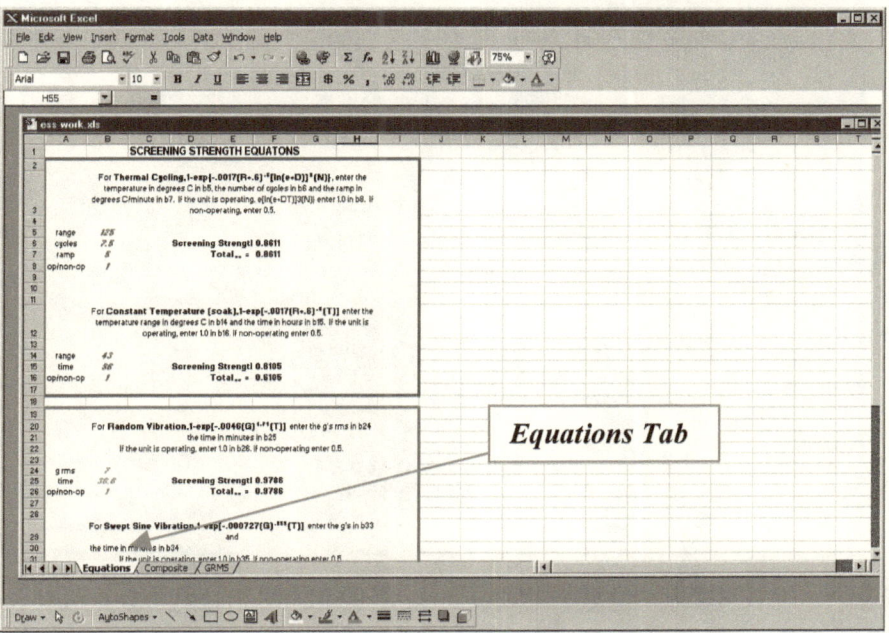

SCREENING STRENGTH EQUATONS

For **Thermal Cycling,** $1-\exp\{-.0017(R+.6)^{.6}[\ln(e+D)]^{3}(N)\}$, enter the temperature in degrees C in b5, the number of cycles in b6 and the ramp in degrees C/minute in b7. If the unit is operating, $e[\ln(e+DT)]3(N)\}$ enter 1.0 in b8. If non-operating, enter 0.5.

range	125	
cycles	7.5	Screening Strength : 0.8611
ramp	5	Total$_{ss}$ = 0.8611
op/non-op	1	

For **Constant Temperature (soak),** $1-\exp[-.0017(R+.6)^{.6}(T)]$ enter the temperature range in degrees C in b14 and the time in hours in b15. If the unit is operating, enter 1.0 in b16. If non-operating enter 0.5.

range	43	
time	96	Screening Strength : 0.6105
op/non-op	1	Total$_{ss}$ = 0.6105

For **Random Vibration,** $1-\exp[-.0046(G)^{1.71}(T)]$ enter the g's rms in b24 and the time in minutes in b25
If the unit is operating, enter 1.0 in b26. If non-operating enter 0.5.

g rms	7	
time	30.0	Screening Strength : 0.9786
op/non-op	1	Total$_{ss}$ = 0.9786

For **Swept Sine Vibration,** $1-\exp[-.000727(G)^{.853}(T)]$ enter the g's in b33 and the time in minutes in b34
If the unit is operating, enter 1.0 in b35. If non-operating enter 0.5.

g's	2	
time	30.0	Screening Strength : 0.0304
op/non-op	1	Total$_{ss}$ = 0.0304

For **Single Frequency Vibration** , $1-\exp[-.00047(G)^{.49}(T)]$ enter the g's in b41, time in minutes in b42
If the unit is operating, enter 1.0 in b43. If non-operating enter 0.5.

g's	12	
time	30.0	Screening Strength : 0.0465
op/non-op	1	Total$_{ss}$ = 0.0093

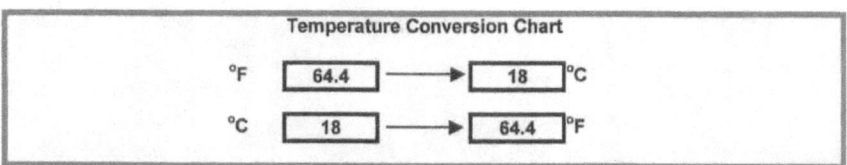

Temperature Conversion Chart

°F	64.4	⟶	18	°C
°C	18	⟶	64.4	°F

A.2 Appendix A – Composite Tab

This excel worksheet can calculate the composite Screening Strength for up to ten (10) screens/variables. Enter the screening strength number in ss1 through ss10 and the "Composite" will be automatically calculated.

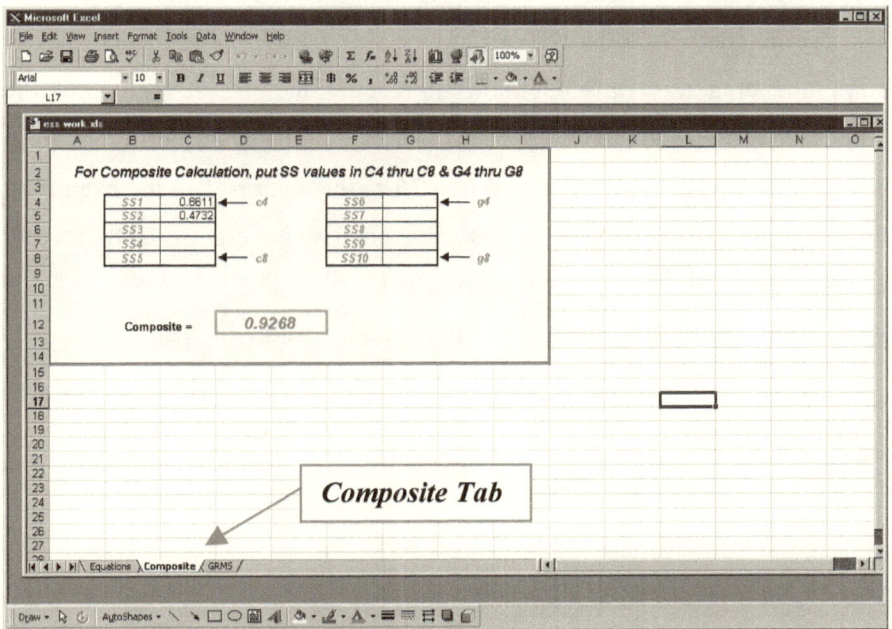

127

A.3 Appendix A – GRMS Tab

Here is an example of calculating 'G RMS" for the vibration equations. This example uses the standard from the Navy P-9492 document. This vibration profile that has been commonly called "NAVMAT". The frequency is from 20 to 2,000 Hz with an amplitude of 0.04 g^2/hz.

The profile starts at 20 hz with an amplitude of 0.01 g2/hz. It increases in amplitude at a rate of +3Db until it reaches 80 Hz, at this point the amplitude has reached 0.04 grms (this is area 1 in the example). From 20 to 350 hz the amplitude remains at 0.04 g2/hz (this is area 2 in the example). From 350 hz to 2,000 Hz the amplitude decreases at a rate of –3Db until an amplitude of 0.007 g2/Hz has been reached.

A printout of the "ESS Work.XLS" file is on the following page

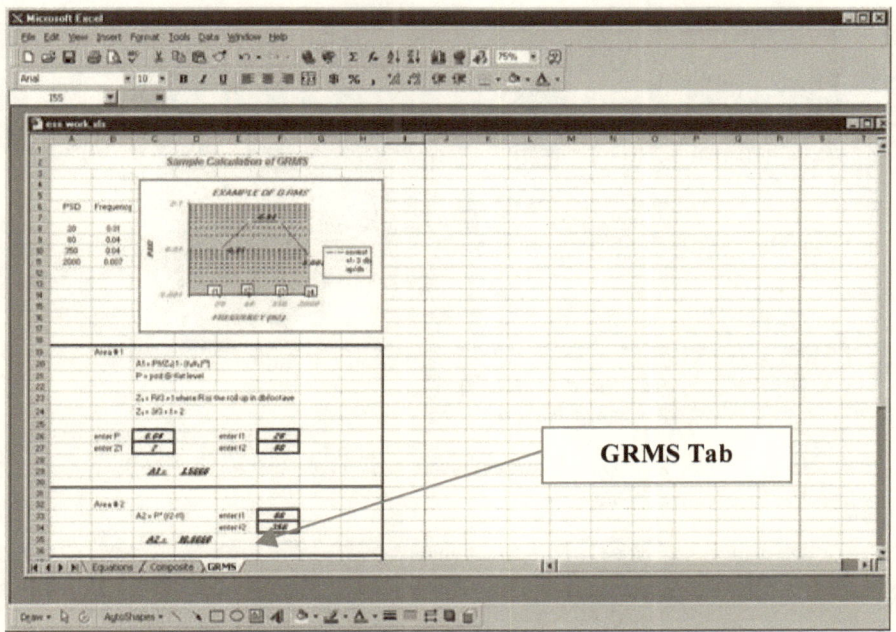

GRMS Tab

John J. Quinn

Sample Calculation of GRMS

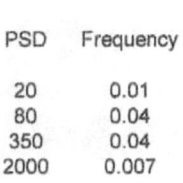

PSD	Frequency
20	0.01
80	0.04
350	0.04
2000	0.007

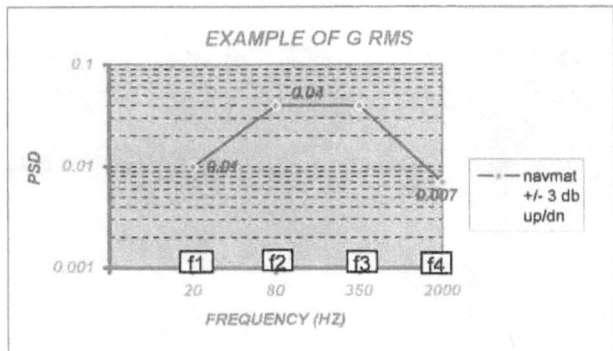

Area # 1

$$A1 = Pf/Z_1\{1 - (f_1/f_2)^{Z_1}\}$$

P = psd @ flat level

Z_1 = R/3 + 1 where R is the roll up in db/octave

Z_1 = 3/3 + 1 = 2

enter P	**0.04**		enter f1	**20**
enter Z1	**2**		enter f2	**80**

A1 = 1.5000

Area # 2

$$A2 = P^* (f2-f1)$$

		enter f1	**80**
		enter f2	**350**

A2 = 10.8000

Area # 3

$$A3 = Pf_1/Z_2\{1 - 1/(f_2/f_1)^{Z_2}\}$$

Z2 = R/3 - 1, where R is the roll down in db/octave

enter P	**0.04**		enter f1	**350**
enter Z2	**3**		enter f2	**2000**

A3 = 4.6417

WARNING: If r = 3, use A3 = 2.3 P f₁ log (f2/f1) =   **24.4016**

Total Area is the square root of A1 + A2 + A3 =   **16.9417**

(r=3) total Area is the square root of A1 + A2 + A3 =   **36.7016**

g rms = 4.1160   **g rms = 6.0582**

(r=3)

Appendix B
Acronyms

John J. Quinn

DOD-HNBK-344 (USAF)

B.1 DEFINITIONS AND ACRONYMS

Assembly – A number of parts joined together to perform a specific function and capable of disassembly, for example a printed circuit board. An assembly of parts designated to function in conjunction with similar or different modules when assembled into a unit. (i.e. Printed Circuit Assembly, Power Supply Module & Core Memory Module).

Defect Density - Average number of Latent defects per item. Symbols used D_{in}, D_{out}, D_r and D_o for incoming, outgoing, remaining and observed defect density respectively.

Detectable Failure – A failure that can be detected with 100 % test detection efficiency (D_e).

Escapes – A portion of incoming defect density which is not detected by a screen and test and which is passed on to the next level. Symbol is (D_{out}).

Failure – Free Period – A contiguous period of time during which an item is to operate without the occurrence of a failure while under environmental stress.

Failure – Free Test – A test to determine if equipment can operate without failure for a predetermined time period under specific stress conditions.

Failure Rate – The total number of failures within an item population, divided by the total number of life units expended by that population during a particular measurement interval under stated conditions. A reliability measure related to MTBF. Symbol is (●).

Fallout – Failures observed during, or immediately after, and attributed to stress screens. Symbol used is (F).

Latent Defect – An inherent or induced weakness, not detectable by ordinary means, which will either be precipitated to early failure under environmental stress screening conditions or eventually fail in the intended use environment.

Part – Any identifiable item within the product which can be removed or repaired (e.g. discrete semi conductor , resistor, IC, solder joint, connector).

Part Fraction Defective – The number of defective parts contained in a part population divided by the total number of parts in the population expressed in parts per million (PPM).

Patent Defect – An inherent or induced weakness which can be detected by inspection, functional test, or other defined means without the need for stress screening.

Precipitation of Defects – The process of transforming a latent defect into a Patent defect through the application of stress screening.

John J. Quinn

Production Lot – A group of items manufactured under essentially the same conditions and process.

Screenable Latent Defect – A latent defect which has a inherent failure rate of greater than 10-3 failures per hour under field stress conditions.

Screen Effectiveness – A measure of capability of a screen to precipitate latent defects to failure. Sometimes used specifically to mean Precipitation Efficiency (Pe).

Screen Parameters – Parameters in Precipitation Efficiency (Pe) equations which relate to Screening Strength, (e.g. vibration g levels, temperature rate of change and time duration).

Screening Regime – A combination of stress screens applied to an equipment, identified in the order of application (i.e. assembly, unit and system screens).

Screening Strength – The probability that a screen will precipitate a latent defect to failure, given that a latent defect susceptible to the screens is present. Symbol is (SS).

Selection and Placement – The process of systematically selecting the most effective stress screens and placing them at the appropriate levels of assembly.

Stress Screening – The process of applying mechanical, electrical and/or thermal stresses to an equipment item for he purpose of precipitating latent part and workmanship defects to early failure.

System/Equipment – A group of units interconnected or assembled to perform some overall electronic function (ie. Electronic flight control system, communications system).

Test Detection Efficiency – A measure of test thoroughness or coverage which is expressed as the fraction of patent defects detectable, by a defined test procedure, to the total possible number of patent defects which can be present. Symbol used is (D_e) used synonymously as the Probability of Detection.

Thermal Survey – The measurement of thermal response characteristics at points of interest within an equipment when temperature extremes are applied to the equipment.

Unit – A self-contained collection of parts and/or assemblies within one package performing a specific function or group of functions, and removable as a single package from an operating system (i.e. auto-pilot computer, vhf communications, transmitter).

Vibration Survey – The measurement of vibration response characteristics at points of interest within an equipment when vibration excitation is applied to the equipment.

John J. Quinn

Yield – The probability that an equipment is free of screenable latent defects when offered for acceptance.

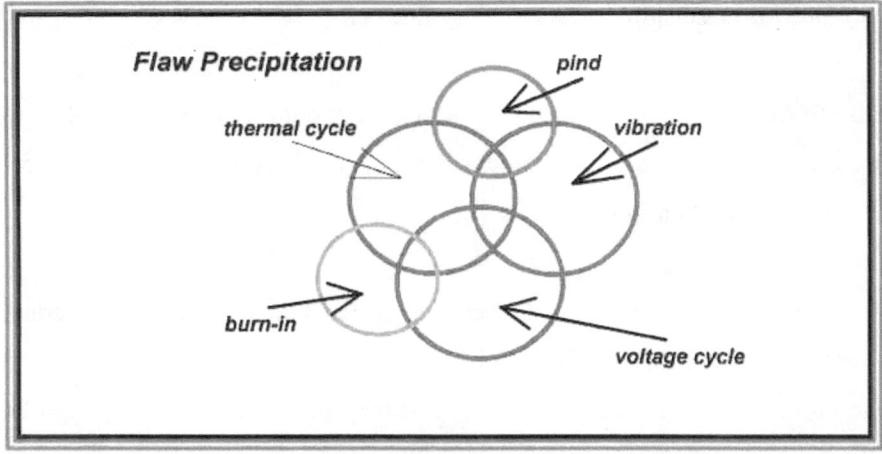

Appendix C
Screening Review Board
&
FRACAS

C.1 SCREENING REVIEW BOARD

A Screening Review Board (SRB) will be comprised of personnel from Manufacturing, PMO, PA, Environmental Engineering, Design Engineering, Reliability Engineering and customer. Formal meeting will be held monthly or as often as required to review, analyze, and assess the effectivity of the ESS processes to determine whether to:

- Continue with the stress screening profiles unchanged
- Increase or decrease screening strength
- Selectively eliminate ESS tests or evaluations
- Initiate new screening steps or procedures
- Redefine original screening goals

Other objectives of the SRB will be to:

- Verify/regulate contractual compliance
- Audit data collection methods/formats
- Recommend screen modifications to PMO & Customer
- Generate monthly customer CDRL reports
- Provide summaries of SRB activities

Specific tasks and responsibilities are as follows:

1)   Manufacturing and Environmental Engineering shall maintain a database for all ESS activity to provide a monthly summary of process yields.

2)   Manufacturing and Environmental Engineering shall generate fault documentation for all ESS failures and trouble shoot any failures in a timely manner.

3)   The SRB will respond immediately to address screening failure issues.

4)   The SRB will review each TFD and assign action items to specific members to perform required analysis to determine failure mechanism and establish the failure category (Manufacturing process, Design, component, or materials).

5)   The SRB will close out all TFDs generated during ESS.

6)   The SRB shall determine if corrective action via Investigation Requests (IR) is necessary to resolve ESS problem issues.

7)   The SRB will review the yield and failure analysis data to assess the effectivity of each screening process and recommend alterations to improve the process.

Corrective action recommended will eliminate the cause of the failure in the most practical and expedient way. In the ESS operation, the recommended action is guided by Table 5.29 of DOD-HNBK-344.

Condition		Comparison Actual vs Planned		Effect on Remaining Dr Goal	Future Scrn costs	Actions Din	Required Screening Strength
		Din	SS				
I	a	hi	lo				
	b	hi	ok	higher than expected	increase	essential	increase Pe
	c	ok	lo		↑	↑	↑
II	d	hi	hi	if higher			
				uncertain			
	e	lo	lo	if lower			
	f	ok	hi		↓	↓	↓
III	g	lo	ok	lower than expected	reduce	by opportunity *	reduce Pe
	h	lo	hi				
IV	i	ok	ok	likely to be achieved	reduce	by opportunity *	no change or eventually reduce
* = Corrective action should always be taken when the opportunity presents itself and costs to take actions are reasonable.							

Table 2 - Comparison of Actual vs. Planned Din and SS Values

C.2 FRACAS

Procedures for hardware test reporting, analysis and corrective action are based upon discipline reporting and computer based information system. The Test Plans/Procedures flow from top level requirements and parameter specifications to the test requirements. Failure Reports (FR) can be generated for all out of specification conditions detected during test. A FR will result from any hardware discrepancy between test requirement parameters and the test results obtained during test. The FR is the primary input to the FRACAS system.

FRACAS Definition

A failure reporting and corrective action system (FRACAS) is used to document failures, analyze failures to determine the root cause and implement corrective action within a closed loop system. FRACAS is also used to distribute data, issue summary reports, and provide data for generating plots and tables to engineering and management.

The fundamental process includes, but is not limited to the following:
1) Failures detection/verification
2) Failure recording
3) Failure analysis
4) Corrective action recommendation
5) Monitor results of corrective action to determine effectiveness

FRACA Process

The detection of the failure is entered into the FRACA database. Quality Assurance reviews the failure data, analyzes the failure and recommends a corrective action. Quality Assurance is always in the feedback loop and is the custodian of the completed Failure Report (FR) and resulting corrective actions.

Time is critical since corrective action in the early maturing phase of the design is much more cost effective than when the system is fully matured and in the field. Accurate information describing the failure is essential for proper failure diagnosis and proper recommended corrective action.

A copy of the FR stays with the defective part, which is routed to Quality Assurance for analysis and determination of the failure mechanism mode.

FAILURE REPORT					389631

Unit Name		model	s/n	date	location
major unit					
assy					
sub assy					
description of test					
description of trouble					
action taken					
summary of analysis					
recommended corrective action					
action taken effectivity					
failure code:					
originator					

Sample – Failure Report

C.3 Rework

Rework has always been a big issue. "How do we handle rework"? "How many times can an item be reworked without damaging it to a point where it has to be scrapped"? Here is a simplified explanation. The rework sequence is shown in the diagram below. There are two paths, one for thermal and one for vibration.

First the thermal sequence. The unit under test (UUT) is repaired and re-tested under ambient conditions to ensure that the repair procedure did not damage the UUT. Three additional thermal cycles are performed to ensure that the UUT can survive the temperature profile. The UUT is tested during the thermal screen if the profile calls for it or during an ambient post test. If the UUT passes, the UUT is returned to the original screening process. If the UUT fails, the UUT is returned to the rework process at point A. At some point the rework process must stop. In this example if the UUT fails three rework attempts, the UUT is sent to a Material Review Board (MRB) for disposition. In MIL-STD-2164 there is more detailed information about recommended thermal levels and time. "It should be noted that, in practical terms, as many additional thermal cycles as are necessary may be applied without affecting the equipment's useful life"(MIL-STD-2164).

Now for the vibration rework sequence. The UUT is repaired and re tested at ambient like the thermal sequence. The UUT is then vibrated using the same ESS profile it failed at but the vibration level is reduced by 6 Db. This is a reduction of approximately one-quarter of the original profile. This will ensure that the UUT does not exceed its life expectancy. The UUT is tested to ensure it is performing to its requirements, as was the thermal sequence. If the UUT passes, it is returned to the vibration screen. If it fails, it is returned to the rework process at point B. Again the 3-fail rule applies. In MIL-STD-2164 there is more detailed information about recommended vibration levels and time. "Unlike thermal cycling, the

John J. Quinn

maximum time that a unit can be exposed to a specified spectrum of random vibration, without significantly affecting its useful life, is severely limited (MIL-STD-2164).

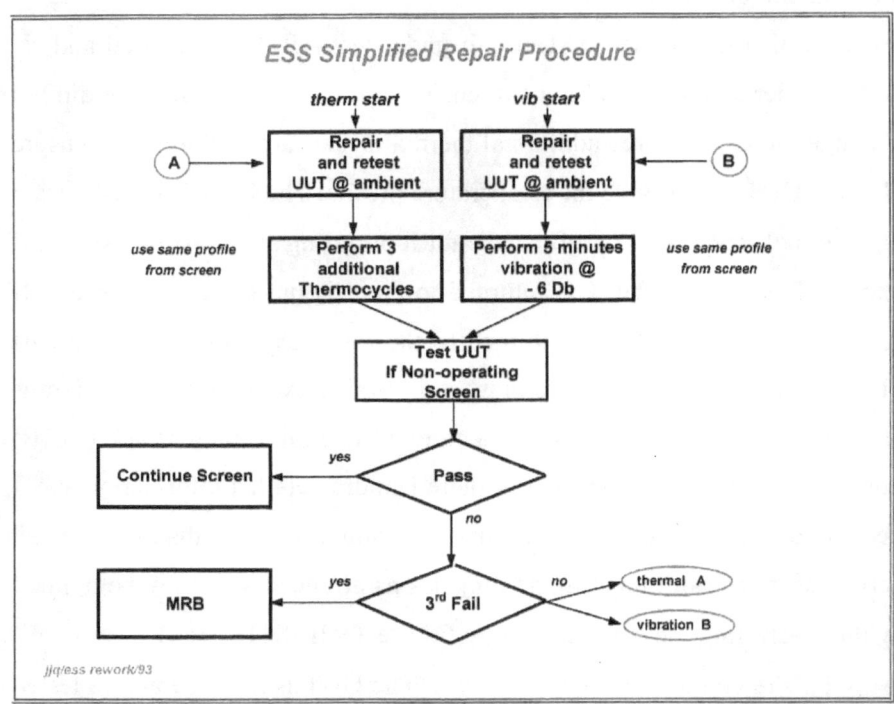

About the Author

Born August 16, 1935, Yonkers ,N.Y.

U.S. Navy, 1952 to 1956, Attended Aerial Photographer "a" school, station at Photo interp center Washington D.C. and USS Leyte (carrier). rate was AF3.

Graduated Westchester college, Valhala, N.Y., 1959, AAS degree.

Field Engineering 1959 to 1961 - worked on first Air Traffic Control System developed in Pleasantville, N.Y. and installed it in NAFEC N.J. The principal hardware worked was a punch and printer that supplied the aircraft information to the aircraft controller's console.

Instructor for Defense contractor on Apollo program, guidance computer. Prepared course content and taught field service engineers and customer support personnel.

Graduated Northeastern University BS, 1969. Attained 3.4 average.

Regional manager for commercial Hospital information system using Programmed Data Processor (PDP) PDP-8 and PDP-11 systems. The systems included automated health history and a Hospital Information System.

Instructor and field support for TRIDENT I program. Developed a training course for field and government personnel for the guidance system (main computer to Inertial Measurement Unit to Flight control computer) and taught course at customer support sites.

1980 - Manager for commercial computer manufacturer, Environmental Test Facility, environments included Temperature/ Humidity, vibration, audio, ESD, FCC requirements for class A and B and TEMPEST (classified testing). The major emphasis was to audit all new and existing electronic products for company compliance. Develop Environmental

Stress Screening (ESS) plans for new products. Scheduled ESS seminars to upgrade company knowledge of ESS.

Presently working for a defense contractor generating ESS and Environmental Qualification Plans/Procedures/Reports.

Major interest in ESS started in the early 1980s while managing an Environmental Audit Facility. It was evident that something had to be done about product being shipped and not working at the customer's site. It worked when it left the factory and now it didn't. I read every document I could find on this problem and it all pointed to Environmental Stress Screening. Hughes Company was the first to analytically approach the problem. They developed algorithms using temperature and vibration during the manufacturing cycle. This finally led to the government developing a document called DOD-HNBK-344. I present a modified approach in this book.